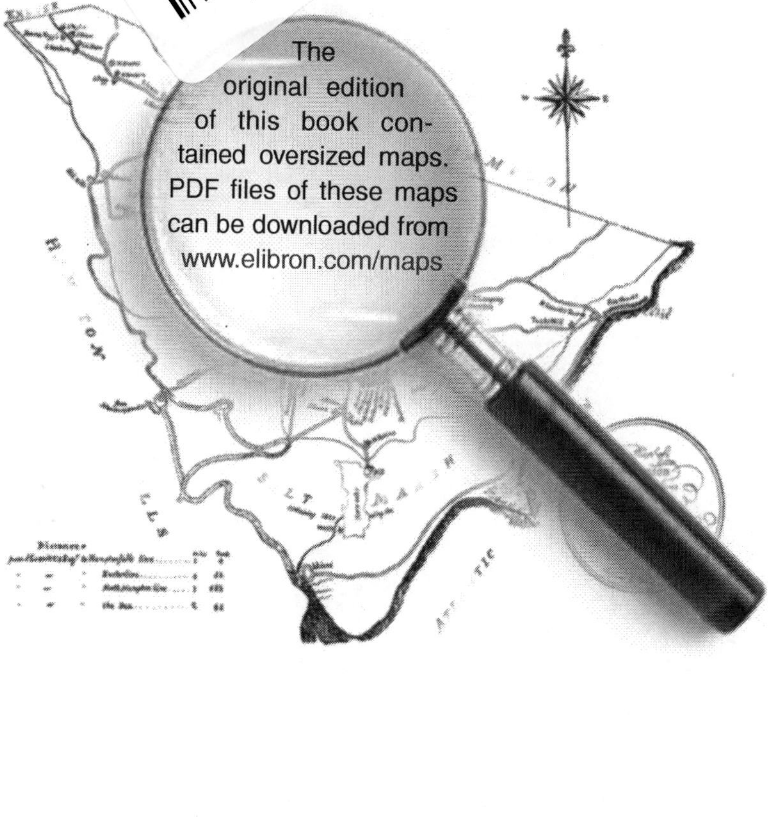
The
original edition
of this book con-
tained oversized maps.
PDF files of these maps
can be downloaded from
www.elibron.com/maps

SHERARD OSBORN

THE DISCOVERY

OF

A NORTH-WEST PASSAGE

BY

H.M.S. INVESTIGATOR, CAPT. R. M'CLURE

DURING THE YEARS

1850-1851-1852-1853-1854

Elibron Classics
www.elibron.com

Elibron Classics series.

ISBN 0-543-98278-5 (paperback)
ISBN 0-543-98277-7 (hardcover)

This Elibron Classics Replica Edition is an unabridged facsimile of the edition published in 1865 by William Blackwood and Sons, Edinburgh and London.

THE DISCOVERY

OF

A NORTH-WEST PASSAGE

THE DISCOVERY

OF

A NORTH-WEST PASSAGE

BY

H.M.S. INVESTIGATOR, CAPT. R. M'CLURE

DURING THE YEARS

1850–1851–1852–1853–1854

EDITED BY

CAPTAIN SHERARD OSBORN, C.B.

ROYAL NAVY

FROM THE LOGS AND JOURNALS OF CAPTAIN ROBERT LE M. M'CLURE

FOURTH EDITION

WILLIAM BLACKWOOD AND SONS
EDINBURGH AND LONDON
MDCCCLXV

NARRATIVES

OF

VOYAGE AND ADVENTURE

BY

SHERARD OSBORN, C.B.

CAPTAIN ROYAL NAVY

IN THREE VOLUMES

VOL. II.

THE DISCOVERY OF A NORTH-WEST PASSAGE

BY H.M.S. INVESTIGATOR

WILLIAM BLACKWOOD AND SONS

EDINBURGH AND LONDON

MDCCCLXV

TO

THE RIGHT HONOURABLE

THE LORDS COMMISSIONERS OF THE ADMIRALTY

THESE PAGES,

DESCRIPTIVE OF THE VOYAGE OF

HER MAJESTY'S DISCOVERY-SHIP INVESTIGATOR

WHILE IN SEARCH OF THE

EXPEDITION UNDER CAPTAIN SIR JOHN FRANKLIN, R.N., K.C.B.

AND OF THE DISCOVERY OF A NORTH-WEST PASSAGE

BY CAPTAIN ROBERT LE MESURIER M'CLURE, R.N.

ARE, WITH THEIR LORDSHIPS' PERMISSION,

Respectfully Dedicated

BY THEIR LORDSHIPS' MOST OBEDIENT AND

HUMBLE SERVANT,

SHERARD OSBORN.

PREFACE TO THE SECOND EDITION.

I HAVE no desire to take unto myself any credit for this work having so rapidly run through its First Edition; but it is a source of no small gratification to find that the discipline, endurance, and gallantry evinced by British officers and seamen under no ordinary trial, which I have endeavoured to chronicle for the honour of my profession, are so warmly appreciated by our countrymen.

My gallant friend Captain F. L. M'Clintock had placed me under deep obligations for the kind manner in which his valuable observations upon the fauna of the Arctic Archipelago were made available. They are embodied with my own in a new chapter.

To Sir Roderick Murchison my thanks are also due, on behalf of the Investigators as well as from myself, for his valuable papers of General Remarks upon the Geological Specimens and Fossils brought home by Captain Sir Robert M'Clure.

My opinions upon the abandonment of a more recent expedition in the arctic regions have been mistaken for those of Captain Robert M'Clure. I have therefore erased them from this work—the more willingly as it has been shown to me that the record of the greatest arctic achievement of our day can be rendered perfect without connecting it in any way with the saddest tale in naval history.

LONDON, *January* 1, 1857.

PREFACE TO THE FIRST EDITION.

THE annals of arctic history afford so many noble illustrations of the spirit of enterprise and hardihood of our sailors, that they will, it is to be hoped, never fail to interest the British people. Of course it is easy to attempt to cast ridicule on any generous impulse of a nation or an individual, by speaking of it as Quixotic, foolhardy, and so forth; but if it be a weakness in English seamen, that for three centuries they have sought to win honour and renown in regions where the ordinary hardships of those whose business is upon the great waters are multiplied a hundredfold, it will assuredly be no joyous day for England, when her sailors shall be free from the charge of any such chivalrous extravagance.

Sir John Franklin and his hundred and thirty-eight gallant followers went forth to achieve the North-West Passage. They discovered it, and perished victims to their zeal. They were followed by one worthy to follow in their footsteps—Captain Sir Robert Le Mesurier M'Clure; he came, indeed, too late to save Franklin; but at least he thoroughly completed the search for him on one given line, by passing from ocean to ocean, and he secured to the Royal Navy and to Great Britain the imperishable renown of having successfully accomplished an enterprise long attempted in vain.

The Editor feels that, in the following narrative, he has scarcely done justice to the many noble qualities of every individual forming the gallant company of the Investigator; but he has at any rate endeavoured to place on record some feeble acknowledgment of their heroic courage and self-devotion.

The delay in the production of this work arose from the Editor having been obliged to leave England upon active service during the Russian war; and the many calls upon his time and attention have rendered perhaps still more imperfect the naturally unpolished style of a sailor's narrative. But into the accuracy of this narrative the Editor challenges the closest investigation; for his ambition has been that this work may remain as the history of a great event in naval chronicles, and perhaps awaken in the breasts of future Franklins, Parrys, or

M'Clures that love for perilous adventure, which must ever form the most valuable trait in the character of a maritime people.

The Editor has not indulged in wholesale praise, for it was no part of his task to write up every man a hero who sailed into the arctic seas and out again. Indiscriminate commendation is loathsome to all right-minded men; and it would be poor reward indeed to those whose tale of suffering and gallantry is recounted in the following pages, to compare their successes with the failures as rife in the arctic seas as elsewhere.

For information on various points, and for assistance in the pleasing but anxious task of collating this narrative with various authorities, the Editor has been indebted to Captain (now Sir Robert) M'Clure, Commander Gurney Cresswell, John Barrow, Esq., F.R.S., Captain Washington, Hydrographer to the Admiralty, John Hay, Esq., of the Admiralty, and other kind friends, to whom he tenders his hearty thanks.

H.M.S. Medusa, Sea of Azov,
April 1856.

CONTENTS.

PAGE

CHAPTER IX.

CHAPTER X.

CHAPTER XI.

CHAPTER XII.

CHAPTER XV.

CHAPTER XVI.

A List of the Officers and Crew of H.M.S. Investigator who performed the North-West Passage.

Name.	Rank or Rating.	Remarks.
R. J. Le M. M'Clure, .	Commander.	
Wm. H. Haswell, .	Lieutenant.	
Samuel G. Cresswell, .	Do.	
H. H. Sainsbury, .	Mate.	Died on board H.M.S. Resolute, off Cape Cockburn, Barrow Strait, 14th Nov. 1853.
Robert Wynniatt, .	Do.	
Stephen Court, . .	Second Master.	19th April 1853, rated Acting Master.
Alex. Armstrong, M.D.,	Surgeon.	
Henry Piers, . .	Assistant-Surgeon.	
Joseph C. Paine. .	Clerk in charge.	
George J. Ford, . .	Carpenter.	
George Kennedy, .	Acting Boatswain.	
Richard A. Ross, .	Quartermaster.	24th Dec. 1850, disrated A.B.
John Davies, . .	A.B.	15th April 1853, rated Quartermaster.
John Kerr, . . .	Gunner's Mate.	Died 13th April 1853, at Bay of Mercy, Banks Land, on board H.M.S. Investigator.
Henry Bluff, . .	Boatswain's Mate.	
Samuel Mackenzie, .	A.B.	
Charles Steel, . .	A.B.	
David Harris, . .	A.B.	
Edward Fawcett, .	Boatswain's Mate.	
James Evans, . .	Caulker.	
George Gibbs, . .	A.B.	
James Williams, .	Captain of the Hold.	
Peter Thompson, .	Captain of the Foretop.	
Samuel Relfe, . .	A.B.	
Thomas Morgan, .	A.B.	Died on board H.M.S. North Star, at Beechey Island, 22d May 1854.

List of Officers and Crew of H.M.S. Investigator—*Continued.*

Name.	Rank or Rating.	Remarks.
John Eames, . .	A.B.	Died 11th April 1853, Bay of Mercy, Banks Land, on board H.M.S. Investigator.
William Batten, .	A.B.	
Charles Anderson, .	A.B.	
Isaac Stubberfield, .	Ship's Cook.	
Frederick Taylor, .	A.B.	
Henry Gauen, . .	Carpenter's Mate.	
George Brown, . .	A.B.	24th December 1850, rated Quartermaster.
Cornelius Hulott, .	Captain's Coxswain.	
William Whitefield, .	Carpenter's Crew.	
Michael Flynn, . .	Quartermaster.	
Mark Bradbury, .	A.B.	
James Nelson, . .	A.B.	
William Carroll, .	A.B.	
George Olley, . .	A.B.	
John Calder, . .	Captain of Forecastle.	
John Ramsay, . .	A.B.	
Henry Stone, . .	Blacksmith.	
Henry Sugden, . .	Sub. Officers' Steward.	
Henry May, . .	Quartermaster.	
Joseph Facey, . .	Sailmaker.	
James M'Donald, .	A.B.	
George L. Milner, .	Gun-room Steward.	
John Wilcox, . .	Paymaster and Paymaster's Steward.	
Robert Tiffeny, . .	Captain of Maintop.	
John Boyle, . .	A.B.	Died 6th April 1853, Bay of Mercy; the first death. Lieutenant Pim arrived next day.
Thomas Toy, . .	A.B.	
Samuel Bonnsall, .	A.B.	
Ellis Griffiths, . .	A.B.	
Mark Griffiths, . .	A.B.	
John Keefe, . .	A.B.	
Thos. S. Carmichael, .	A.B.	
John Woon, . .	Sergeant of Marines.	
J. B. Farquharson, .	Corporal ,,	
George Parfitt, . .	Private ,,	
Elias Bow, . . .	Private ,,	
James Biggs, . .	Private ,,	15th April 1853, rated Corporal.
Thomas Bancroft, .	Private ,,	
Thomas King, . .	Private ,,	
James Saunders, .	Private ,,	
Johan A. Miertsching, .	Esquimaux Interpreter.	

SAILING ORDERS.

From the Lords Commissioners of the Admiralty to Captain COLLINSON, C.B., *of Her Majesty's Ship Enterprise, dated 15th January* 1850.

By the Commissioners for executing the Office of Lord High Admiral of the United Kingdom of Great Britain and Ireland.

1. WHEREAS the efforts that have been made during the last two years to relieve the Erebus and Terror have failed, and all access to the Parry Islands has been prevented by the accumulation of ice in the upper part of Barrow Strait: And whereas it is possible that the same severity of weather may not prevail at the same time in both the eastern and western entrances to the Arctic Sea, we have now determined, in a matter of such moment, to send an expedition into the Polar Sea from the westward; and, having a full confidence in your zeal and skill, we have thought proper to appoint you to the command of Her Majesty's ship Enterprise, and also to place under your orders Her Majesty's ship Investigator; both of which vessels having been duly fortified against collision with the ice, equipped for the polar climate by warm-air apparatus, and furnished with provisions for three years, as well as a large supply of extra stores, you are now required and di-

rected, so soon as they are in all respects ready for sea, to proceed to make the best of your way to Cape Virgins, in order to arrive at Behring Strait in July.

2. At Cape Virgins, the Commander-in-Chief in the Pacific has been desired to have a steam-vessel waiting for you, and by her you will be towed through the Strait of Magellan and the Wellington Channel, and on to Valparaiso.

3. At that port you will use the utmost despatch in watering and refreshing your crews, and in fully replenishing your bread and other provisions and stores; and having so done, you will again use your best exertions to press forward to the Sandwich Islands.

4. There is only a bare possibility of your reaching those islands in time to meet Her Majesty's ship Herald, under the command of Captain Henry Kellett; but if that should be the case, you will receive from him not only every assistance, but much useful information touching your passage to the Strait, and your further proceedings to the northward. It is still more improbable that Her Majesty's ship Plover should be there; but wherever you may fall in with her, you are hereby directed to take her and Commander Moore under your orders.

5. At the Sandwich Islands you will find additional orders from us for your guidance, which we propose to forward from hence by the Panama mail of next March; but if none should arrive, or if they do not in any way modify these directions, you will enforce the greatest diligence in re-victualling your two vessels, in procuring, if possible, the necessary Esquimaux interpreters, and in making all requisite preparations for at once proceeding to Behring Strait, in order to reach the ice before the 1st of August.

6. An examination of the several orders issued to Captain Kellett, will show that it is uncertain where he may be fallen in with. You may probably find the Herald and Plover together.

7. We consider it essential that, after entering the ice, there should be a depot, or point of succour, for any party to fall back upon. For this purpose the Plover is to be secured in the most favourable quarter, as far in advance as can be found —such as Wainwright Inlet, or the Creek at Hope Point; but if they be unsafe, and none has been discovered nearer to Barrow Point, then at Chamisso Island, or any part of Kotzebue Sound, which may afford the necessary shelter.

8. Considering, however, the nature of the service in which the Plover will already have been employed, and that a portion of her crew may be unfit to contend with the rigours of a further stay in those latitudes, you will call for volunteers from that ship, and from the Herald, if in company, sufficient to form a crew for the Plover; taking care that the men to be selected are men of good character, and that they do not exceed in number what is actually required for the care of the ship, and for defence and security against any treacherous attack on the part of the natives of Norton Sound.

9. The petty officers' ratings that may be vacated by men invalided are to be filled up by men volunteering to remain; such volunteers are to be subjected to a strict and careful survey by the medical officers of the several ships; and those only are to be retained who would seem to be in all respects fit to encounter this extended service; and the remainder necessary to complete the crew is to be made up from the Enterprise and Investigator.

10. Such crew having been formed (to continue under the command of Commander Moore, and with the officers now in the Plover, or with those who may volunteer for the service), the Plover, if the Herald should be in company, is to be filled up by Captain Kellett with all the provisions, fuel, and stores that can possibly be spared by Captain Kellett, who will bear in mind not only what may be required for the use of the Plover's crew until the autumn of 1853, and the contingency of parties arriving on board from Sir John Franklin's expedi-

tion, but also the possibility of any party from the Enterprise or Investigator having to fall back upon the Plover.

11. In providing for this necessary equipment for the Plover, attention will be paid to the numbers left in the Herald, and the supplies necessary to carry that vessel to Whoahoo; and having received from Captain Kellett any baidars, or light boats, that he may be able to spare, and which may be likely to form a useful addition to your own boats, or those of the Investigator, when searching-parties may be detached from the ships in the spring, the Herald will return to the Sandwich Islands, there to fill up provisions, and from thence proceed to Hong-Kong on her way to England, in pursuance of our orders of the 14th December last.

12. On detaching the Plover to take up her winter-quarters, you will direct Commander Moore to remain there until you join him, or failing your return to him, until the end of the summer of 1853; when, but not until it is absolutely necessary for securing the Plover's passage through the Aleutian group of islands, he is to quit Behring Strait, and make the best of his way to Valparaiso (touching at the Sandwich Islands for refreshment), where he will receive further instructions relative to his return to England from the Commander-in-Chief.

13. If the Herald and Plover should be fallen in with to the northward and eastward of Behring Strait, or in the Polar Sea, Captain Kellett, on detaching himself from your company, should consort with the Plover as far as her winter-quarters, and if time and circumstances admitted of it, he should assist in securing her there.

14. In the event of your having to winter your ships on the continent or Esquimaux shores, you will probably meet with some of the wandering tribes, or with Indians. With these you will cultivate a friendly feeling, by making them presents of those articles to which they are apt to attach a value;

but you will take care not to suffer yourself to be surprised by them, but use every precaution, and be constantly on your guard against any treacherous attack. You will also, by offering rewards, to be paid in such manner as you may be able to arrange, endeavour to prevail on them to carry to any of the settlements of the Hudson Bay Company an account of your situation and proceedings, with an urgent request that it may be forwarded to England with the utmost possible despatch.

15. In whatever place you may have to establish your winter-quarters, you will devote every resource in your power to the preservation of the health, the comfort, and the cheerfulness of the people committed to your care.

16. We leave it to your judgment and discretion as to the course to be pursued after passing Point Barrow, and on entering the ice; and you will be materially assisted in this respect by what you will learn from Captain Kellett, if he should be fallen in with at the Sandwich Islands, as well as from the observations of Sir E. Parry and Captain Beechey contained in the memoranda, of which we send you copies.

17. We have desired that you shall be furnished, not only with a copy of the orders under which Commander Moore is now acting, but also with copies of all the orders which from time to time have been given to Captain Kellett, as well as with those under which an attempt was made to relieve the Erebus and Terror by Captain Sir James Ross on the eastern side through Baffin Bay. You will further be supplied with all the printed voyages or travels in those northern regions; and the memoranda and instructions drawn up by Sir John Richardson, as to the manners and habits of the Esquimaux, and the best mode of dealing with that people (a copy of which is also sent), will afford a valuable addition to the information now supplied to you.

18. We deem it right to caution you against suffering the two vessels placed under your orders to separate, except in

the event of accident or unavoidable necessity; and we desire that you will keep up the most unreserved communication with the commander of the Investigator, placing in him every proper confidence, and acquainting him with the general tenor of your orders, and with your views and intentions from time to time; so that the service may have the full benefit of your united efforts in the prosecution of such a service; and that in the event of any unavoidable separation, or of any accident to yourself, Commander M'Clure may have the advantage of knowing, up to the latest period, all your ideas and designs relative to the satisfactory completion of this undertaking.

19. We also recommend that as frequent an exchange may take place as conveniently may be of the observations made in the two ships; that any information obtained by the one be as quickly as possible communicated for the advantage and guidance of the other.

20. In case of any irreparable accident happening to the Enterprise, you are hereby authorised to take command of the Investigator, and make such arrangements for the officers and crews as may be most consonant to the rules of the service, and most conducive to the objects of the expedition.

21. In the event of Great Britain being involved in hostilities with any foreign power during your absence, you are to abstain from the smallest act of aggression towards any vessel belonging to such nation, it being the practice of all civilised countries to consider vessels engaged in service of this kind as exempt from the rules and operations of war.

22. In carrying out the foregoing orders, you will avail yourself of every practicable occasion of acquainting our Secretary with every step of your progress, as well as with your future intentions; and occasionally during your voyage, you will throw overboard one of the tin cylinders with which you have been supplied (headed up in any cask or barrel that you could manufacture or spare), containing an account of

the date, position, &c. On your reaching England, you will call on every person, in both vessels, to deliver up their logs, journals, charts, and drawings, but which, they may be informed, shall be returned to them in due time.

23. With respect to your search proving fruitless, and your finally quitting the polar seas, as well as your securing your winter-quarters towards the close of any one season, we cannot too strongly impress upon you the necessity of the utmost precaution and care being exercised in withdrawing in time, so as in no case to hazard the safety of the ships, and the lives of those intrusted to your care, by your being shut up in a position which might render a failure of provisions possible.

We feel it unnecessary to give you more detailed instructions, which might possibly embarrass you in a service of this description; we have therefore only to repeat our perfect reliance on your judgment and resolution, both in doing all that is possible to relieve the missing ships, and in withdrawing in time, when you come to the painful conclusion that your efforts are unavailing.

24. You will bear in mind that the object of the expedition is to obtain intelligence, and to render assistance to Sir John Franklin and his companions, and not for the purposes of geographical or scientific research; and we conclude these orders with an earnest hope that Providence may crown your efforts with success, and that they may be the means of dispelling the gloom and uncertainty which now prevail respecting the missing expedition.

Given under our hands, this 15th day of January 1850.

(Signed) F. T. Baring.
(„) J. W. D. Dundas.

By command of their Lordships,

(Signed) J. Parker.

Richard Collinson, Esq., C.B.,
Captain of H.M.S. Enterprise, at Devonport.

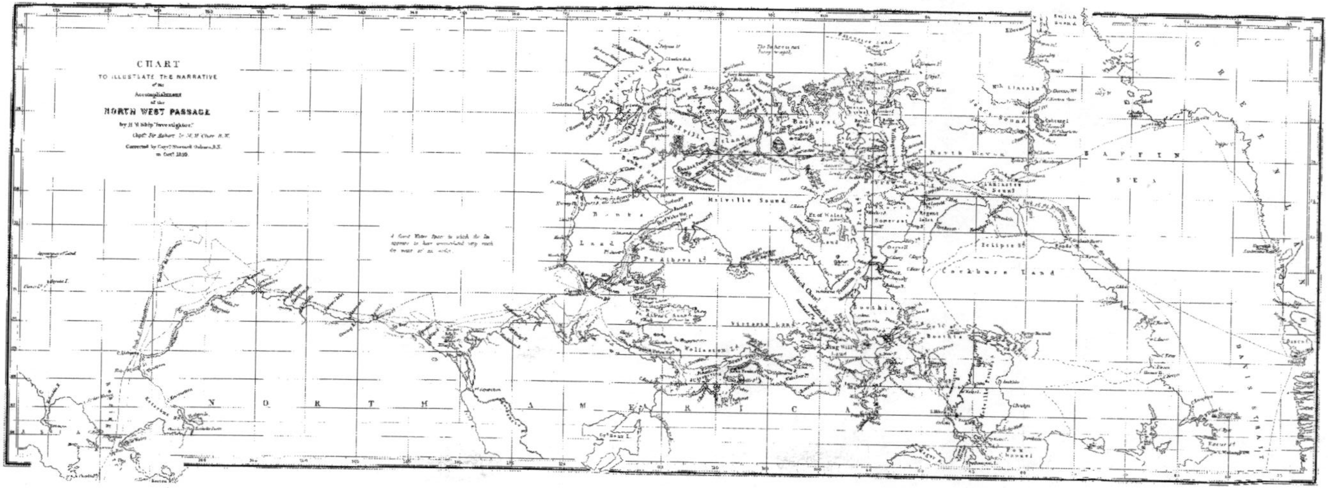
CHART
TO ILLUSTRATE THE NARRATIVE
of the
Accomplishment
of the
NORTH WEST PASSAGE
by H. M. Ship "Investigator."
BAFFIN
SEA
Melville Sound
Banks
Land
Cockburn Land
Eclipse Sd.
North Devon
Jones Sound
Victoria Land
Wollaston Ld.
Melville Island
Bathurst
N O R T H   A M E R I C A
DAVIS STRAIT
GREENLAND

THE DISCOVERY

OF

A NORTH-WEST PASSAGE.

CHAPTER I.

INTRODUCTION.

The successful realisation of the project so long cherished in Great Britain, of the discovery of a way through the Arctic Ocean to the Indies—the final solution of a problem sought through many an arduous struggle during the course of three hundred years—is what it is our present duty to chronicle; and we may be permitted, in the first place, briefly to remind the reader of the reasons that made such a voyage desirable.

On looking at the surface of the globe, it will be seen at a glance that Hindostan, China, and Japan—the Ophirs of the Old World—are placed, geographically speaking, with respect to Western Europe, in the most distant and inaccessible position.

Turn to an ancient atlas, and think of the "antres vast and deserts idle" that lay between India and Europe, and we can better appreciate the forays of the great Macedonian, the difficulties he encountered, and the genius which, in mastering them, raised him to the rank of a demigod among his countrymen. Yet Alexander left no footprints east of the Indus.

The legions of imperial Rome failed to carry their conquering eagles to a region which they must have been fully aware contributed largely to the enormous wealth of Jerusalem, Judea, and Egypt. Even Roman ambition was checked by the difficulties of the route.

The Mohammedan, more fortunate in his central position, served for ages to act as the medium of transit for the spices and products of the East to Western Europe, whither the Crusader carried back a knowledge of and taste for luxuries previously unknown,—a knowledge which created new wants, and excited the mercantile and nautical skill of nations dwelling upon the shores of the Mediterranean. Venice and Genoa rose to greatness upon their lucrative trade with the East; and the fact soon came home to the common sense of their neighbours, and awakened the desire to supplant them, or share in their profits.

In the middle of the thirteenth century, Marco Polo brought back to Western Europe such glowing accounts of the East as verified all the traditionary tales of Cipango and Cathay. Enterprise and cupidity were aroused. The Portuguese slowly but successfully proceeded along that African shore which, as they knew, touched Asia upon the Mediterranean Sea, and which they naturally inferred would eventually lead them to the prize, if they followed it in an opposite direction.

Columbus, with more genius, ascribed to the Indies a vast extent of eastern longitude; and then deduced, from certain traditionary accounts of a land lying west of Iceland (whither he had traded), that by sailing in that direction from the Isle of Fayal he would secure to the prince that should employ him possession of the Indies. Columbus sailed eventually under Spanish auspices, and discovered a land which he fondly imagined was the long-wished-for Eastern continent.

England, not less than Spain and Portugal, was excited to maritime adventure by the wonderful discovery of the Americas, then supposed to be the eastern seaboard of the long-sought Indies. These Indies, in the mean time, were actually reached by the Portuguese, under Vasco de Gama, by way of that Cape of Torments flatteringly named by John II. Cabo da Boa Esperáza.

Before a century had passed, however, both Spaniards and Englishmen found that a wide ocean lay between that Western continent and Cathay; and although the Spaniard could reach the latter by rounding the continent of America at the Strait of Magellan, yet he forbade all others the right to risk the adventure, and treated as pirates all of our countrymen who dared approach the Great South Sea. The Portuguese, in like manner, monopolised the route round the Cape of Good Hope, and both added to the danger that awaited the English navigator in his efforts to share the booty of the East.

John Cabot, who, in the service of Henry VII., had

secured for Britain a title to American discovery, left a son, Sebastian, whom Edward VI. wisely patronised. With this man seems first to have originated the idea of reaching the coveted land by taking a north-about route through polar seas, and thus avoiding the dangers of the great sea voyage by either of the southern capes.

Our seamen, accustomed to the hardships and cold of northern latitudes, feared the frigid as little or less than the torrid zone; and just forty years * after Nunez de Balboa waded into the sea at Panama, and, striking it with his sword, claimed the broad Pacific as the property of Spain, the first English arctic expedition sailed to achieve a *north-east* passage to Cathay.

On May 10, 1553, three ships left Greenwich, under Sir Hugh Willoughby, and were rewarded by discovering Novaia Zemlia and the White Sea; but the subsequent melancholy fate of the admiral and his crew, who were frozen or starved to death in Lapland, checked the national enthusiasm; and although two more attempts were made in the same direction, they were equally unsuccessful. The Dutch, like ourselves, anxious to share in the wealth of the Indies, so jealously watched by Spaniards and Portuguese, tried to reach them by a north-east route: their countryman Barentz perished in gallant and unavailing attempts to carry out their wishes; and after him England made two slight efforts in the same direction under Hudson and Cherie, and then turned her attention in the opposite one. Prior to the time of Barentz's last north-eastern expedition, the first

* Vasco Nunez de Balboa discovered and took possession of the Pacific Ocean in 1513.

north-west voyage had already been attempted from Great Britain.

Martin Frobisher, a seaman of Queen Elizabeth's time, was the first to call attention to the possibility of reaching the East, Cipango, and Cathay, by passing north of America. For fifteen years he patiently urged his ideas upon the consideration of his countrymen, and at last, in 1576, in the reign of Elizabeth, he sailed to commence that work which the seamen of England only completed in 1854, or two hundred and seventy-four years afterwards, in the reign of Queen Victoria.

It is not our purpose to follow Frobisher, Davis, Hudson, and Baffin, through their adventures and discoveries. Our charts of the present day, on which the localities they discovered bear their respective names, sufficiently attest the amount of the additions they made to our geographical knowledge; but when the last-named great navigator, Baffin, returned in 1616 from the most successful of his voyages, the North-West Passage was still, as Martin Frobisher said, "a thing yet undone, wherebye a notable mind might be made famous and fortunate."

Years elapsed—men's minds took another turn—times changed—the maritime supremacy of Britain gradually developed itself. If Spain could boast of her Magellan, England had had her Drake.

The tide of fortune now flowed slowly in favour of our hardy race of seamen. Along the seaboard of Spanish America, whether in the Atlantic or Pacific Oceans, the English seamen slowly but steadily established their pre-eminence; and within a brief period from the date of the first Indian cargo * being exposed in Leadenhall Market, a footing was obtained in India, and the foundation of our eastern empire firmly established.

From 1632, when Captains Fox and James discovered those northern localities by the names of which their memories have still been preserved, a century elapsed before the attempt to discover a north-west passage to India was resumed. Even then, although a reward of £20,000 was offered, the attempts of Scroggs, Dobbs, and Middleton, in 1740-41, yielded no better fruit than many of our modern voyages,—a vast deal of writing, exaggerated accounts of difficulties, and no results of any value.

Captain Cook's success in the South Seas, and, indeed, wherever he went, led the nation to hope that he might be the man fated to secure to his country the honour of a discovery which was then desired on commercial as well as geographical grounds. He failed, however, in penetrating the ice, and well was it for himself and his crews that it was so; the fate of Sir Hugh Willoughby would assuredly have befallen them, unprepared as they were for such a voyage, and the rigours of such a climate.

About that time two of the servants of the Hudson Bay Company reached the mouths of the Mackenzie and Coppermine rivers, and sighted an arctic sea lying north of the American continent; but the year 1800 found us still far from the accomplishment of an enterprise com-

* On the 15th of September 1592 the first cargo of Indian produce was exposed in Leadenhall. It was captured by George Clifford, Earl of Cumberland, in the Portuguese carrack Madre de Dios, off the Azores, and was sold as prize property.

menced more than two hundred years before. Small results were to be seen on our charts for such long and patient toil, suffering, and devotion to their duty as the majority of the navigators engaged in arctic discovery had exhibited.

On the side of the Pacific, by Behring Strait, the termination of our knowledge of the American continent was at Icy Cape; and on the side of the Atlantic, at Hudson Bay. Between this space of eighty degrees of longitude, the arctic sea had only been seen at two points. The rest was a blank. Immediately after the ratification of the general peace of Europe in 1815, the idea of carrying into effect an achievement, upon which so much zeal and wealth had already been expended, was revived with great earnestness; and it is evident, in the memoirs of Haines Barrington and Sir John Barrow, that at the outset they cherished the most ardent hopes of securing to their country a discovery that would be both useful and glorious.

At this juncture a Russian expedition under Kotzebue, which had been equipped at the expense of Count Romanzoff, threatened to rob us of the prize. England's pride was aroused, and the 'Quarterly Review' of January 1818 expresses it in the following words:—"It would be somewhat mortifying," says the writer (the late Sir John Barrow, F.R.S.), "if a naval power but of yesterday should complete a discovery in the nineteenth century, which was so happily commenced by Englishmen in the sixteenth, and another Vespuccio run away with the honours due to a Columbus."*

* 'Quarterly Review,' January 1818, page 219.

In 1816 a more than usual concurrence of favourable winds, currents, and weather had disengaged and carried down a vast body of ice from the polar regions north of the Atlantic. Icebergs and floes were found floating in great quantities as far south as the 40th parallel of latitude. The very climates of both northern Europe and America were affected by the decreased temperature they occasioned; and, as an instance of it, Indian-corn would not ripen in 1816 in either Pennsylvania or Massachusetts, —an unparalleled circumstance.

Mr Barrington, as well as Sir John Barrow, both maintained that, could a navigable route be found north-about, the commerce of England with the East, more than that of any other nation, would be benefited; and the higher the latitude in which we could pass into the Pacific, the more the distance would be shortened. "For instance," says the latter, "the distance from Shetland to Behring Strait, in the 72d parallel, is just half as long again as on a meridian passing through the pole, or as 1572 leagues to 1048."

The people of England entered into these views, and into all the theories and conjectures of Sir John Barrow. Some of these conjectures are quite startling to one who, like the writer, is able to sit down and peruse them, knowing on what slender premises they were grounded, and seeing how many have actually been verified since his death.

In October 1817, Sir John Barrow published a small diagram to illustrate an article of his upon the existence of a north-west passage, which is now before us; and although he was only then in possession of the information which we have said England possessed at the end of the eighteenth century, yet, guided by a clear judgment and a thorough knowledge of the subject, he filled up the deficient coast in so correct a line that the charts of to-day, upon the same small scale, vary but little from his.

The search for a north-west passage was now actively resumed, and voyage after voyage followed rapidly—Ross, Parry, and Franklin executing, and Barrow cheering them on.

Captain James Burney, who, as a lieutenant of Captain Cook, carried some authority with his opinion, opposed the idea of any communication between the Pacific and Atlantic Oceans, and actually connected the American and Asiatic continents.

The first navigator chosen to follow on the footsteps of Baffin attempted to close every avenue or outlet to the north-west from Baffin Bay. Yet Sir John Barrow was not daunted.

"I have every reason to believe Old Greenland an island, or an archipelago of islands," he said, "and no inclination to deny that some of them may stretch far enough to the westward to form those several sounds of which Baffin so briefly and vaguely speaks." Shortly after the above lines were written, Parry verified them by really sailing west through an *archipelago of islands* half way to Behring Strait!

Sir John Barrow, the great promoter of arctic research, while sharing the honours of every fresh discovery, was certainly not desirous of avoiding the responsibility attached to sending men upon a service of so much risk and danger, and was ready to adduce good reasons for doing so. Until the repeated failures of Parry and Franklin convinced him of the impossibility of using a north-west passage for commercial purposes, Sir John Barrow had evidently fondly hoped to secure to his country the advantages of a shorter route to India: but even then, jealous of her maritime glory, and influenced by a sincere love of science and a desire for the extension of knowledge, he judged rightly that nowhere could the skill and energy of the British navy be more honourably directed than to geographical discovery, whether in the frozen or torrid zones.

Arctic exploration had, indeed, entailed some hardship and suffering; but be it remembered that up to the day when Sir John Franklin sailed on his last voyage in 1845, no fatal catastrophe had overtaken any one of the many ships that had been employed on that service in modern days.

Sir John Barrow appreciated fully the difficulties and dangers of arctic service; and a mere selfish desire to see knowledge extended, at any sacrifice of humanity, was assuredly not his failing. He did not live to see the realisation of the confident predictions of his sagacious mind; yet he had seen a vast deal added to our geographical knowledge of the earth's surface—additions of no small moment in every branch of natural history, as well as in physical science; and the wide world saw and knew that the race of seamen who had secured to Britain the supremacy of the sea in war, were not deteriorating in time of peace.

It was no narrow or selfishly ambitious feeling that

dictated the following words, during the time that one expedition was absent in search of the long-sought passage. "They may not succeed," he says, "in that purpose; but they can scarcely fail in being the means of extending the sphere of human knowledge, and if they bring back an accession of this, they cannot be said to have been sent in vain, for 'knowledge is power,' and we may safely commit to the stream of time the beneficial results of its irresistible influence."

Franklin and his hundred and forty followers were the forlorn hope of the North-West Passage. By the sacrifice of their lives, they have secured to us, their countrymen, an honour that perhaps might otherwise never have been won; for it was in seeking for them that Captain M'Clure and his gallant officers and crew succeeded, for the first time in the annals of the world, in passing from the Pacific to the Atlantic Ocean. In the eloquent words of Lord Stanley and Sir Edward Parry, when addressing Captain Cresswell, who first brought the intelligence to England, in 1853, of the discovery of the water communication between the Pacific and Atlantic Oceans, "it was a triumph that would not be valued the less highly because it was not stained by bloodshed—a triumph which was not embittered by any single painful or melancholy reminiscence—a triumph not over man, but over nature—a triumph which inflicts no injury, and which humiliates no enemy—a triumph not for this age alone, but for posterity—not for England only, but for mankind."

When such is the testimony borne to the honour won by those who had achieved this passage, Sir John Barrow needs no apologist for having been the main promoter of arctic research; and although the laurel is twined with the cypress-wreath of those who have laid down their lives in this service, their friends and relatives may proudly wipe away their tears.

The North-West Passage would never have been discovered but for the devotion of Franklin, his officers, and men; they each volunteered for that duty, and they fell in the performance of it. The party from the Erebus and Terror, which perished, we now know, at the mouth of the Great Fish River, went down the channel which leads from Capes Walker and Bunny in Barrow Strait, and they, thus dying, forged the great link which connects the known coasts of the Parry Archipelago with that of the American continent. They did not, like the crew of the Investigator, achieve the passage by actually passing from ocean to ocean; but it is possible that at the very moment when Captain M'Clure stood on the northern coast of Banks Land, and assured himself of a water communication between the Pacific and Atlantic Oceans, some lonely survivor of Franklin's expedition might have been watching from King William Land, that known highway to Behring Strait, which they well knew extends from King William Land and the Great Fish River. Captain M'Clure and his followers can well afford to surrender cheerfully to the illustrious dead that share of the honour which is their due; and we who mourn the loss of those who perished with Franklin in gallantly endeavouring to perform what the Investigator so happily effected, may still point to the chart of the polar regions, and say, "*Si monumentum requiris, circumspice!*"

CHAPTER II.

RETURN OF THE ARCTIC SQUADRON OF 1848, AND IMMEDIATE RE-EQUIPMENT OF THE ENTERPRISE AND INVESTIGATOR—APPOINTMENT OF CAPTAIN M'CLURE—PREPARATIONS FOR SPEEDY DEPARTURE—THE TWO VESSELS PASS THE NORE—HEAVY GALES IN THE CHANNEL—PUT INTO PLYMOUTH—ENGAGEMENT OF AN INTERPRETER—FINAL DEPARTURE FROM ENGLAND—APPEARANCE OF ARCTIC SHIPS—INVESTIGATOR REACHES THE SOUTH ATLANTIC.

THE Arctic Discovery Squadron, under Captain Sir James Ross, had hardly returned from its perilous operations of 1848 and 1849, when it was at once determined by Government to re-equip the vessels, in order that another expedition might resume the search after Sir John Franklin's missing ships by the way of Behring Strait.

The Enterprise and Investigator, it will be remembered, had failed in their attempt to get to the westward of Leopold Island in 1849, and only escaped from that inhospitable spot, to be beset in the drifting pack-ice of Barrow Strait, to be swept with it out of Lancaster Sound into Baffin Bay, and thence just secured their retreat to England before the arctic seas became generally sealed for that season.

Tempest-tossed and ice-worn though the good ships were, our naval dockyards soon put them into proper condition for once more resuming their contest with floe and iceberg. Captain Richard Collinson, C.B., was appointed to the Enterprise as senior officer of the expedition, and Commander Robert Le Mesurier M'Clure to the Investigator. The former officer enjoyed a high naval reputation, and in China his abilities as a surveyor had done the State good service; the latter, the destined discoverer of a north-west passage, had served through a long and severe probation in every grade, until, after a struggle of twenty-six years, he was appointed to the command of the Investigator, as a reward for the good service he had rendered as first-lieutenant to Sir James Ross, in his voyage of 1848-49.

There was in the winter of 1849-50 no lack of volunteers for arctic service amongst the officers and men of the Royal Navy. The most sanguine feelings concerning the cause they were about to engage in, animated the whole service, and told with excellent effect in the speedy equipment of the ships and the completion of their crews; and although the pendants had been hoisted as late as the 19th of December 1849, yet, on the 10th of January following, the Enterprise and Investigator were enabled to put to sea from the river Thames with their full complement of men. They were, however, much hampered upon both upper and lower decks with provisions and stores; and heavy winter gales in the Channel having caused both vessels to strain much, it became advisable to put into Plymouth to be caulked—a measure which gave the captains an opportunity for entering several more good seamen from that well-known nursery of men-of-war's men.

No grass was allowed to grow under their feet at Plymouth, for the vast distance between England and Behring Strait had to be traversed by way of Cape Horn, in deep-laden bluff-bowed ships ;—winter gales awaited them in the Channel, and equinoctial ones off Cape Horn. All knew there was a weary six months' voyage before they could reach the ice, and that the loss of a month by accident or neglect might cause the highway they sought from Behring Strait to Melville Island to be closed against them.

At last nearly all was ready; every article of equipment on board, and what was not then procurable was ordered to be sent to the Sandwich Islands, via Panama. We must not fail to mention that, before sailing, there was shipped on board the Investigator a German clergyman, a Mr Mierching, who had been engaged as an Esquimaux interpreter. The proposal had been suddenly made to the worthy man one day when he was enjoying his ease in a quiet village in Saxony, after returning from a long sojourn in Labrador as a Moravian missionary. He accepted the offer, and was despatched forthwith as fast as rail could take him to London. The Admiralty sent him a few hours afterwards by express to Plymouth, and he arrived only just in time to be tumbled into the Investigator before her departure.

On the 20th of January 1850 this arctic squadron weighed, and the Enterprise and Investigator sailed thence with a fair and fresh wind. It will now be my duty to follow the latter vessel and her gallant company in their long and adventurous voyage, at the same time avoiding minute details of the everyday operations of the Investigator, which partake rather of the character of a ship's log than of remarks intended for the general reader.

I must endeavour to picture to the general and non-professional reader the appearance of the little vessel, whose name will be remembered as long as England holds dear the character of her adventurous arctic navigators.

Many of our readers have seen in olden times, during westerly winds in the British Channel, water-washed looking vessels of four or five hundred tons register rolling towards the Thames, freighted with rich produce from the East or West Indies. These craft could boast of little beauty, and had but little speed. People said they were fine *wholesome* vessels. Very likely they were; at any rate, they took a long time to get through their work, and looked as if they were built by the mile, and cut off in lengths, as required for the trade. Such a one was H.M.S. Investigator, for she had been bought from an eminent firm at Blackwall. The "fast sailing copper-bottomed A 1, &c.," had been doubled with wood in every direction; and her bow and stern, according to then infallible rules for arctic ships, were made to resemble the ends of a caisson, by having wood and iron bolted one over the other, until in some places twenty-nine inches of solid timber might have been found, or even more. Everything in the shape of outward ornament was of course carefully eschewed; and a solemn coating of black paint, but little relieved by a white riband and small figure-head, added yet more to the appearance of strength and weight of the little, but I am afraid I cannot say *pretty*, vessel.

Aloft, in her powerful rigging and large blocks, were seen signs of a departure from the perfect neatness of a Queen's ship—for appearance had to be sacrificed to efficiency; and her deeply-immersed hull, her decks covered with casks securely lashed, and the many strange articles, such as sledges, ice-triangles, ice-saws, and crow's nest, all told their own tale, and were well in keeping with her character and destination. The crew of this good ship was, as the sequel will show, not unworthy of her, for throughout Captain M'Clure's journal I meet constant expressions of admiration for the men under his command; indeed, so early as the 24th of January, he says, when speaking of the accidental loss of his topmast:—"Vexatious as it certainly was, still it was attended with one advantage, as it gave me an insight into the disposition of my crew, and a more orderly set of men have seldom been collected."

Thus early was that mutual confidence sown which afterwards yielded such good fruit.

Battling with foul winds and heavy seas, it was not until the 20th of February that the Investigator reached the N.E. trade-wind of the Atlantic. Whilst passing through the tropics, care was taken during the fine weather of that region to make good such defects as had shown themselves in the vessel, and were within the power of her artificers to remedy; and all due precautions were employed to insure the health and comfort of officers and men.

CHAPTER III.

SLOW SAILING OF THE ARCTIC SHIPS—REACH TERRA DEL FUEGO—H.M.S. GORGON IN POSSESSION BAY—REACH PORT FAMINE—SHIPWRECKED AMERICAN SCHOONER—AMERICAN GO-AHEAD-ISM—SUBLIME SCENERY—INVESTIGATOR MEETS THE ENTERPRISE—AMERICAN RIVER-STEAMER IN THE PACIFIC—HEAVY GALE—CAPTAIN M'CLURE'S CARE OF THE MEN.

It was not until the 18th March, nearly two months after leaving England, that the Investigator passed out of the southern tropic in the South Atlantic Ocean, although a heavy press of canvass had constantly been carried. It was nearly a month afterwards that Captain M'Clure sighted Cape Virgin, that headland of the South American continent which marks the eastern entrance of the Strait of Magellan. This rate of progress gives the best idea of the speed of the arctic discovery-ships, and best enables us to appreciate the anxiety with which Captain M'Clure, his officers, and crew, must have watched the precious hours as they flitted by, as well as the credit they deserve for the perseverance and ability with which they did so much, with such inferior means.

On the 15th of April, the land of the Strait of Magellan gladdened their sight, and, inhospitable as it was, it was welcome after eighty-five days of sea work;

and they commenced beating along that coast where so many points bear English names strangely mixed up with the Spanish ones—names that tell how England's gallant seamen have toiled, and hoped, and suffered in the great cause of human advancement, from Anson and Drake, proud admirals, to poor Allan Gardner, the sailor missionary, who perished in a noble effort to carry civilisation to the wretched Fuegians. On Point Dungeness were seen large troops of guanachoes, a species of the llama: their flesh would have been a great acquisition, but a lee shore gave Captain M'Clure enough to do to keep his ship in safety, until upon the turn of tide he was enabled to push westward against the contrary wind which prevails in this strait. Next day the Investigator reached Possession Bay, and found H.M.S. Gorgon there, ready to assist in towing her into the Pacific Ocean; and accordingly, by her aid and the zealous exertions of her commander, the Investigator, on the 17th April, arrived at Port Famine.

At this wretched Chilian penal settlement, Captain M'Clure learnt that the Enterprise had already passed, and, what was still more to be regretted, taken with her all the bullocks, so that the prospect of fresh beef for the Investigator's crew was no nearer than the Sandwich Islands, to reach which the wide Pacific Ocean had to be traversed as the Atlantic had already been. The most interesting object seen at Port Famine was the castaway crew of an American schooner, bound to California. The stoical indifference to the reverse of fortune with which it had pleased Providence to visit the two Yankee owners —for they were there—struck our arctic explorers much;

and the unshaken firmness with which they determined not to be beaten was illustrative of the best phase of American character.

Their tale was a strange, but not an unusual one. In the height of the Californian fever then raging, one of the owners had been a vendor of hardware, the other the proprietor of a dry provision store in New York. The latter, walking one day on the quay, saw a schooner advertised for sale; and being struck with the number of persons daily embarking for California, the thought suddenly entered his head that in that El Dorado he might make a fortune far more rapidly than even in the smart town of New York. At this moment his eye chanced to light upon the vendor of hardware, likewise apparently in a brown study; and addressing him, the dry-goods-man asked, "what he thought of a Californian venture?" "Just what I was thinking of," was the reply; and with true American *go-aheadism*, they at once decided on selling off, and embarking their all, four thousand dollars each, in the schooner. The thing was done; and ten weeks afterwards they were cast away upon the south extreme of America, and utterly ruined!

Yet they were not dispirited, and they wanted no sympathy. "No!" said the dry-provision merchant, when it was suggested that he had better return to New York and adhere to his old line of business—"no! I guess I'll never go back to New York; I'll get somehow to California, and right myself yet!"

The Gorgon took some of the men on board until they could be put into their own countrymen's vessels.

Leaving Port Famine, the Investigator proceeded on

her way, passing the wild and, in many places, sublime scenery which skirts the shores of this wonderful and intricate inland sea—the lofty mountains presenting at once the characteristics of the equatorial and polar regions, their summits covered with glaciers, whilst their bases are clothed with dense and humid forests whose rank luxuriance of vegetation resembles that of the thickest Indian jungle.

After rounding Cape Frowards, they found the Enterprise at anchor in Fortescue Bay, and had a good opportunity of comparing their relative courses since parting company. By the charts, the two discovery-ships had seldom been sixty miles apart, and they had crossed the equator within thirty-five miles of each other, though without meeting. Such are the not unusual accidents of a sea voyage; and at any rate neither Captain Collinson nor his colleague could accuse themselves of being the cause of delay to the other. Whilst at this anchorage, detained by strong and foul winds, an American river-steamer, bound to California, passed by. She had many passengers, and took our voyagers' mail to Valparaiso, it being the intention of her skipper, a reckless sort of individual, to take the in-shore channels which lead from Cape Horn to Chiloe, before he trusted his frail-built vessel to the mercy of the Pacific.

That steamer appeared to have made such a marvellously quick passage, that some of the Investigator's officers naturally asked the skipper if he were certain of his dates of departure and arrival. The reply, if not pleasing to ears polite, contained in it a great fact: "Well, officer," said Jonathan, "I ain't sure, for I have lived so long amongst liars that I don't know when I am speaking the truth!"

On the 19th of April the weather moderated sufficiently for the Gorgon to take both the Enterprise and Investigator in tow, and they started at an early hour against a breeze which otherwise would have still kept sailing ships idle. Fuegians, in frail canoes, chased them in hopes of barter, but were unable to communicate, and next day the vessels passed Cape Pillar, the western boundary of the strait.

Once in the broad Pacific, the swell of a western sea caused the hawsers to carry away so often as to prevent the towing of more than one vessel at a time. The Enterprise was first taken, and a good offing given her, and then the Investigator was helped on in her turn. As night came on, however, the gale freshened from the north-west, and the Enterprise and Investigator parted company, never again to rejoin.

During the night of the 20th, the Gorgon was obliged to cut away the towing-hawser of the Investigator, and after lying by her until daylight, Captain Paynter, of the Gorgon, bore up to look for the Enterprise.

So heavy a gale now came on that no canvass could be carried by the Investigator to bear her away from the iron-bound shores of Patagonia, distant at the time not more than thirty or forty miles; and thus drifting before an increasing gale, she was driven far to the south-west, and it was not until the 30th of April that it abated, and enabled the tempest-tossed ship to resume her course for the next rendezvous, Honolulu, in the Sandwich Islands.

At this early period of the voyage, every precaution was taken by Captain M'Clure to prevent the seeds of scurvy being sown amongst his men. He increased the rations of vegetables and fresh preserved meat, issued them every alternate day, and, as soon as the weather permitted it, the officers and crew were constantly employed in their watches restowing, examining, and ascertaining the quality of those provisions, upon which Captain M'Clure's experience told him the result of his voyage would so much depend. No one knew better than he that in the far north man is a machine not to be replaced; that without the zealous labour and the skill and energy of the crew the best officers are worthless. To work a willing seaman to death, or into scurvy, and then turn round and lay the blame on the climate of the poles, was not the system of the man who was about to achieve the North-West Passage; and in adopting the course he did, he took the best method of convincing his men, that, so far as circumstances admitted, he would always study their health and comfort. The crew on their side naturally repaid him with unbounded confidence and hearty exertions.

CHAPTER IV.

VOYAGE CONTINUED TO HONOLULU—LEAK IN THE BREAD-ROOM—LONELINESS OF THIS OCEAN—GOOD FEELING BETWEEN OFFICERS AND MEN—ARRIVAL AT HONOLULU—REPLENISHING AND DEPARTURE—GREAT ANXIETY OF OFFICERS AND MEN TO REACH THE ICE—PASSING THE ALEUTIAN ISLANDS—DENSE FOGS—THE ARCTIC CIRCLE CROSSED—MEET THE PLOVER DEPOT-SHIP—UNFAVOURABLE REPORT OF THE STATE OF THE ICE—CAPTAIN M'CLURE'S PLAN OF OPERATIONS—PREPARATIONS FOR MEETING THE ICE—H.M.S. HERALD MET WITH—CAPTAIN KELLETT'S DISCOVERY.

THE whole of the month of May 1850 was passed by the Investigator in making the best of her way towards Honolulu; and the only incident worthy of notice was the loss of a topmast, and the discovery of the destruction of a large quantity of biscuit, owing to a leak which had broken out in the bread-room: nearly a thousand pounds of biscuit were found to be so mouldy that it had to be thrown away; but Captain M'Clure consoled himself with the certainty of being able to replenish his stock at the first port.

On 2d June they passed the limit of the S.E. trade-wind in lat. 17° 44′ S., long. 105° 54′ W. During the fortnight that the Investigator had been steering diagonally across that space which intervenes between the southern tropic and the equator of the Pacific Ocean, monotony pressed heavily upon both crew and officers. Five weary months of salt water is enough to try the patience of any men; and Captain M'Clure's journal here laments the loneliness of the vast region of water which rolls from America to the Polynesian Isles, a fine and fair wind, clear sky, and smooth sea, hardly compensating for the absence of every living thing except themselves: not even a bird or a fish broke the dreariness of the ocean solitude. Every day served, however, to unite more strongly the bond of mutual good feeling between the commander and his crew. Much heavy work connected with the restowage and examination of the provisions fell, of course, upon the latter; but they did it with such cheerfulness and goodwill as to draw forth the remark in Captain M'Clure's handwriting: "I have much confidence in them. With such a spirit what may not be expected, even if difficulties should arise?" This good opinion of his men was fully borne out in the sequel.

The equator was crossed on the 15th June, and the S.E. trade kindly favoured them into 7° north latitude. On the 23d of June the N.E. trade-wind reached them, and aided by it they made on the 24th the longest run the Investigator had as yet performed in twenty-four hours—namely, one hundred and eighty-six miles. It was a proud day for the old ship.

On June 29th the snow-capped peak of Mouna-Kea, on the island of Owhyhee, showed itself, and announced that at last they were nearing a port, after a sea-voyage of 15,000 miles.

Running through the western islands of the Sandwich group, they anchored, gladly enough, on the 1st July, outside the reefs of Honolulu harbour, the wind being foul for entering its narrow mouth.

There was no time to be lost, however, and all necessary supplies were speedily purchased, Captain M'Clure being much aided in his labour by Captain Aldham of H.M. brig Swift, then in the port. The very next day, all the stores were ready for shipment, and the Investigator was prepared for the prosecution of her voyage with the utmost despatch. The crew of the Swift generously undertook this duty while the men of the Investigator were allowed all the leave that could be afforded them to wander about the beautiful island of Oahu.

Captain Collinson, in the Enterprise, had already, it appeared, called at this port, and had replenished and gone on, in prosecution of his instructions. The anxiety of Captain M'Clure to overtake his chief, and rejoin him in time for entering the polar sea in 1850, can be better conceived than described. Abundant supplies of fruit and vegetables were purchased, although the high prices occasioned by so large a demand, as well as the arrival of the annual fleets of American whalers, formed a considerable obstacle to the private purchases of officers and men. By a remarkable fatality, too, their hopes of obtaining a good supply of fresh beef were again destined to be disappointed; for the natives, in attempting to land the bullocks from a coaster, had drowned all but one of them.

The Consul-General, Miller, as well as some of the merchants, extended due hospitality to Captain M'Clure; and although he had occasion to remark that Yankee influence was all-powerful in these islands—the American stepping-stones to the empire of the East, as they have been appropriately called—yet neither he, nor the Investigators generally, had any cause to complain of want of courtesy towards themselves, or of sympathy for the noble mission on which they were employed.

The 4th July 1850 saw the Investigator in every respect ready for the polar voyage, which was her appointed task. Every available space was again filled with provisions, and in addition to a solitary bullock, a dozen sheep, and vegetables sufficient to last forty days, were embarked. Her weak or sickly men had been invalided, and their places supplied by volunteers; and all hands were much refreshed and reinvigorated by even the short stay they had made in so pleasant a spot. On leaving Honolulu, the same evening, Captain M'Clure remarks that the health and cheerfulness of all were most gratifying, and that no vessel could have entered the ice under more favourable circumstances after so long and trying a sea-voyage.

The ice, however, was still full forty degrees of latitude distant, the Enterprise far ahead, and there were only about sixty days more of summer left. Well, therefore, might Captain M'Clure feel anxious when he looked at his deeply-laden vessel, and pondered on the quickest mode of reaching Behring Strait. The old-established course from the Sandwich Islands to the Strait was by hauling out to the N.W. across the N.E. trade, so as to strike the Asiatic coast in or about the latitude of Petropaul-offskoi, and then taking advantage of westerly winds to run along that coast, and thus avoid the dangerous channels through the Aleutian Islands, which were imperfectly known, and subject to dense fogs and strong tides. This route, however, generally occupied sixty days, and Captain M'Clure had heard of another and more direct one; but at the same time the risk of his vessel being driven down upon the dangerous and inhospitable coast north of Vancouver Island and Princess Charlotte Island was duly pointed out. Fortunately, he had met at Honolulu with an intelligent merchant sailor, who urged him by all means to run the risk, steer a *direct* course to the northward, and not to fear north-west gales at that season of the year. This advice, together with the emergency of the case, induced him to decide upon making direct for the Aleutian Islands, and trusting to the "sweet little cherub that sits up aloft" for the winds he might require.

At midnight a course was shaped accordingly, and M'Clure went to his bed and dreamt of catching the Enterprise; for there was much fear among both officers and men lest a report, heard at Honolulu, should prove to be true, that Captain Collinson intended, instead of waiting for his consort, to take the Plover on into the ice, and leave the Investigator to occupy her place in Kotzebue Sound as a mere depot-ship.

No other bad effects had ensued from the liberty given to the Investigator's men at Honolulu, after their long imprisonment on board, than that six of them were laid up for a few days, in consequence of having, sailor-like, taken a great deal too much equestrian exercise—one or two having actually cut their feet with the stirrup-irons through riding hard without any shoes.

Favoured by strong and fair winds, the ship made

rapid progress to the northward, although dense fogs gave the captain and master much anxiety, lest they should be carried east or west of that channel through the Aleutian Isles, for which they were endeavouring to steer.

On the sixteenth day after leaving Oahu, while expecting to sight the land, the mist lifted sufficiently to show a bluff point, with a detached rock lying off it. This satisfied them that the island they had wished to sight was the one now at hand, and, as the Investigator was swept rapidly along in a tide-race with a fair wind, they had only time to observe, on the bow, the western extreme of the island of Tchunam. It was but a momentary glimpse, however, that they were blessed with of either, and this was all they saw of the Aleutian chain; but when certain of being clear and to the north of the islands, the Investigator was steered for Gore Island.

Some idea may be formed of the denseness of the fogs which prevail in this region, and add so materially to its dangers, from the circumstance of some of the sea-birds, such as the little auk (*Alca alle*), striking against the rigging in their flight, and falling on board the ship.

But fog or no fog, the ship was still pushed on with the utmost press of sail that any regard to safety would permit; yet a sea but little known was before the Investigator,—a sea interspersed with islands whose position was so uncertain that, in some cases, they were not even placed in the chart. Many an anxious hour was passed on board the ship, their greatest fear being lest, by grounding the vessel and detention, they should be too late for entering the ice in the season of 1850.

On 26th July the stud-sails were for the first time taken in since leaving Oahu,—a proof of how happily the winds had hitherto favoured the ship,—and next day they made King Island, and had a narrow escape of being swept on shore by a rapid and unexpected current. Fortunately, a good bearing, by compass, of the land was obtained before they were again enveloped in fog, so that they were able to shape a course between the two Diomede Islands; but still, running the gauntlet, as the Investigator was doing, was fraught with peril, and nothing but the urgency of the case induced Captain M'Clure to persist, although it might be that in such a sea, and amongst so many tides and currents, a straight course on end at high speed was the safer one.

The words of her commander convey the best idea of the ship's position and his feelings as they approached the channel between the Diomede Islands:—"The channel is a good one doubtless in clear weather; but in a fog which has never given us a horizon more than four hundred yards distant, with a very strong and unknown current, with a fresh fair wind and deep water, it was an exceedingly anxious time for me, more especially when an unusual ripple was perceived, resembling the tidal race of Alderney or Portland. The noise was so great that we could not hear what was said without great vocal exertion; the sea was breaking into the channels, and the deep-sea lead showed that the ship was sweeping through twenty-two fathoms' water only." The look-out man even reported breakers ahead; and not small was the general delight when, by the deepening of the soundings and the cessation of the tide-ripple, the voyagers guessed themselves to have been swept into Kotzebue Sound, though disagreeably close past one of the islands at its entrance.

On 29th July 1850 the arctic circle was crossed, and the Admiralty clothing supplied by the Lords Commissioners of the Admiralty for the use of seamen in those regions was issued to the crew. Arctic sights now rapidly accumulated to interest the many novices in the Investigator, and to awaken in the mind of their commander recollections of his former trying and unsuccessful voyage on the opposite side of America into that same frozen sea.

The first to greet them was the Plover depot-ship, then commanded by Commander Moore, who, with two boats, was making inquiries amongst the natives of Wainwright Inlet concerning some of the thousand and one Esquimaux fictions that then enlivened the monotony of Kotzebue Sound. The eager query of how long since the Enterprise had gone to the northward? was answered by the Plover informing them that she had not been fallen in with at all; this the officers of the Investigator presumed might be accounted for from the dense fogs which had lately prevailed having prevented her being seen. The Enterprise had left Honolulu as early as the 30th June; she was the faster-sailing ship, and there was no reason to suppose she had not been equally favoured in winds.

The news of the state of the ice in Behring Strait was most unsatisfactory, especially to those who did not know that its movements are so uncertain that a report more than twenty-four hours old was worthless. The Plover had seen the polar ice so densely packed from Wainwright Inlet upon the American coast to lat. 71° 30′ N., and long. 164° 28′ W., as apparently to defy all efforts to push a vessel to the north-east.

In recording this information, Captain M'Clure adds in his journal this remarkable sentence, showing how decidedly he had made up his mind to the course to be pursued, that of entering the ice at all hazards: "This," says he, "was not what I anticipated from the continuance of south-east winds; but the season is not far advanced, and much may yet be accomplished *ere we are frozen up*."

The Investigators were generally struck with the healthy, and, in most cases, robust appearance of the Plover's crew. Indeed, this was not to be wondered at, considering they had merely wintered on the verge of the arctic zone, and in a place like Kotzebue Sound, where the natives are able to subsist upon the resources of the country, and where monotony, darkness, and hunger could not weigh upon them in any extraordinary degree.

Before bearing up for Cape Lisburne (the next rendezvous that Captain Collinson had appointed), and where the Herald, Captain Kellett, was known to be cruising, all the letter-bags of the Investigator were sent on board the Plover; and at the same time that admirable despatch from Captain M'Clure to the Admiralty, in which he clearly stated what his intentions were, should he be left to act upon them. How to the very letter he carried them out, aided by his gallant officers and men, will be seen in the sequel.

Copy of a Letter from Commander M'Clure to the Secretary of the Admiralty.

"Her Majesty's Discovery Ship Investigator, at Sea
(Lat. 51° 26′ N., Long. 172° 35′ W.),
20th July 1850.

"SIR,—As I have received instructions from Captain Collinson, C.B., clear and unembarrassing (a copy of which I enclose), to proceed to Cape Lisburne, in the hope of meeting him in that vicinity, as he anticipates being detained a day or two by the Plover in Kotzebue Sound, it is unnecessary to add that every exertion shall be made to reach that rendezvous; but I can scarcely venture to hope that, even under very favourable circumstances, I shall be so fortunate as to accomplish it ere the Enterprise will have rounded that Cape, as, from her superior sailing, she has hitherto beaten us, by eight days to Cape Virgins, and from Magellan Strait to Oahu by six. It is, therefore, under the probable case that this vessel may form a detached part of the expedition, that I feel it my duty to state, for the information of the Lords Commissioners of the Admiralty, the course which, under such a contingency, I shall endeavour to pursue; and I have to request that you will lay the same before their Lordships.

"1st. *After passing Cape Lisburne, it is my intention to keep in the open water, which, from the different reports that I have read, appears, about this season of the year, to make between the American coast and the main pack, as far to the eastward as the 130th meridian, unless a favourable opening should earlier appear in the ice, which would lead me to infer that I might push more directly for Banks Land, which I think it is of the utmost importance to thoroughly examine.*

"In the event of thus far succeeding, and the season continuing favourable for further operations, it would be my anxious desire to get to the northward of Melville Island, and resume our search along its shores, and the islands adjacent, as long as the navigation can be carried on, and then secure for the winter in the most eligible position which offers.

"2d. In the ensuing spring, as soon as it is practicable for travelling parties to start, I should despatch as many as the state of the crew will admit of, in different directions, each being provided with forty days' provisions, with directions to examine minutely all bays, inlets, and islands towards the N.E., ascending occasionally some of the highest points of land, so as to be enabled to obtain extended views, being particularly cautious, in their advance, to observe any indication of a break-up in the ice, so that their return to the ship may be effected without hazard, even before the expenditure of their provisions would otherwise render it necessary.

"3d. Supposing the parties to have returned (without obtaining any clue of the absent ships), and the vessel liberated about the 1st of August, my object would then be to push on towards Wellington Inlet (assuming that that channel communicates with the polar sea), and search both its shores, unless, in so doing, some indication should be met with to show that parties from any of Captain Austin's vessels had previously done so, when I should return and endeavour to penetrate in the direction of Jones Sound, carefully examining every place that was practicable. Sir, should our efforts to reach this point be successful, and in the route no traces be discernible of the long-missing expedition, I should not then be enabled longer to divest myself of the feeling, painful as it must be to arrive at such a conclusion, that all human aid would then be perfectly unavailing, and therefore, under such a conviction, I would think it my duty, if possible, to return to England, or, at all events, endeavour to reach some port that would insure that object upon the following year.

"4th. In the event of this being our last communication, I would request you to assure their Lordships that no apprehension whatever need be entertained of our safety until the autumn of 1854, as we have on board three years of all species of provisions, commencing from the 1st September proximo, which, without much deprivation, may be made to extend a period of four years, as, moreover, whatever is killed by the hunting parties I intend to issue in lieu of the usual rations, which will still further protract our resources.

"It gives me great pleasure to say that the good effects of the fruit and vegetables (a large quantity of which we took on board at Oahu) are very perceptible in the increased vigour of the men, who at this moment are in as excellent condition as it is possible to desire, and evince a spirit of confidence and cheerfulness of disposition which are beyond all appreciation.

"5th. *Should difficulties, apparently insurmountable,*

encompass our progress, so as to render it a matter of doubt whether the vessel could be extricated, I should deem it expedient, in that case, not to hazard the lives of those intrusted to my charge after the winter of 1852, *but, in the ensuing spring, quit the vessel with sledges and boats, and make the best of our way to either Ponds Bay, Leopold Harbour, the Mackenzie, or for the whalers, according to circumstances.*

"Finally. In this letter I have endeavoured to give an outline of what I wish to accomplish (and what, under moderately favourable seasons, appears to me attainable), the carrying out of which, however, not resting upon human exertions, it is impossible even to surmise if any or what portion may be successful. But my object in addressing you is to place their Lordships in possession of my intentions up to the latest period, so, as far as possible, to relieve their minds from any unnecessary anxiety as to our fate; and having done this —a duty which is incumbent on me, from the deep sympathy expressed by their Lordships, and participated in by all classes of our countrymen, in the interesting object of this expedition—I have only to add that, with the ample resources which a beneficent Government and a generous country have placed at our disposal (not anything that can add to our comfort being wanting), we enter upon this distinguished service with a firm determination to carry out, as far as in our feeble strength we are permitted, their benevolent intentions. — I have, &c.,

"(Signed) ROBERT M'CLURE, Commander."

Not a letter written that day contained any expression but that of a hopeful issue to their enterprise; and if anxiety was expressed, it was only that of being delayed, or being too late. The calm and resolute spirit of their leader is marked in the paragraph penned this same night. "I consider," he writes, "that we have said adieu to the world for the next two years. May that arm which has conducted us so far in safety, still continue its protection upon a service where all else is weakness indeed!"

On 31st July the vessel was prepared for falling in with the ice; the crow's-nest was sent up to the masthead, ready for the look-out men to take their station in; whale-lines and ice-anchors were placed at hand, ready for heaving, or tracking the ship through loose ice; and ice-chisels, saws, ladders, and all the many articles of equipment peculiar to arctic service, were placed on deck.

The current had swept the Investigator thirty miles north of her reckoning, and Cape Lisburne was far astern when H.M.S. Herald, Captain Kellett, hove in sight.

This vessel, it will be remembered, was annually ordered from her surveying service in Central America, to communicate with, and replenish the provisions of the Plover depot-ship. The Herald usually arrived in June, and left Behring Strait in September. Her object in cruising about the strait was mainly in the hope of falling in with the squadron of Sir John Franklin, should either of his ships have accomplished the passage from the Atlantic to the Pacific Ocean; but the Herald likewise did good service in correcting the charts of this neighbourhood, and in adding materially to our geographical knowledge. The most important discovery, however, that Captain Henry Kellett had made, and one which, in connection with the nature of the ice met with by the Investigator westward of Banks Land, is deeply interesting to those curious about the yet unknown regions which lie around our pole, was the sighting of an extensive land north and north-west of Behring Strait. In Captain Kellett's despatches to England, bearing date 1849, the discovery is graphically described.

"At 3 A.M. the 17th of August, the temperature of the sea suddenly fell from 40° to 36°; the wind became light, and excessively cold. Shortened sail, supposing that I was very near the ice; frequent snow-showers.

"At 5 A.M. wind shifted suddenly from the N.W. in a sharp squall with heavy snow. Shortly after 8, when one of these snow-storms cleared off, the packed ice was seen from the mast-head from S.S.W. to N.N.W., five miles distant. The weather was so bad that I bore up for the rendezvous. The weather, however, as suddenly cleared up, and I hauled my wind for the north-western extreme of the ice that had been seen. At 9.40 the exciting report of 'Land, ho!' was made from the mastheads; they were both soon afterwards crowded.

"In running a course along the pack towards our first discovery, a small group of islands was reported on our port beam, a considerable distance within the outer margin of the ice.

"The pack here was not so close as I found it before. Lanes of water could be seen reaching almost up to the group, but too narrow to enter unless the ship had been sufficiently fortified to force a hole for herself.

"These small islands at intervals were very distinct, and were not considered at the time very distant.

"Still more distant than this group (from the deck) a very extensive and high land was reported, which I had been watching for some time, and anxiously awaited a report from some one else. There was a fine clear atmosphere (such a one as can only be seen in this climate), except in the direction of this extended land, where the clouds rolled in numerous immense masses, occasionally leaving the very lofty peaks uncapped, where could be distinctly seen columns, pillars, and very broken peaks, characteristic of the higher headlands in this sea—East Cape and Cape Lisburne, for example.

"With the exception of the N.E. and S.E. extremes, none of the lower land could be seen, unless, indeed, what I took at first for a small group of islands within the pack-edge was a point of this great land.

"This island or point was distant 25 miles from the ship's track; higher parts of the land seen, not less, I consider, than 60 miles. When we hove-to off the first land seen, the northern extreme of the great land showed out to the eastward for a moment, and so clear as to cause some who had doubts before, to cry out, 'There, sir, is the land quite plain.'

"From the time land was reported until we hove-to under it, we ran 25 miles directly for it. At first we could not see that the pack joined it, but as we approached the island we found the pack to rest on the island, and to extend from it as far as the eye could reach to the E.S.E.

"The weather, which had been fine all day, now changed suddenly to dense clouds and snow-showers, blowing fresh from the south, with so much sea that I did not anchor as I intended.

"I left the ship with two boats; the senior lieutenant, Mr Maguire, Mr Seemann, naturalist, and Mr Collinson, mate, in one; Mr Goodridge, surgeon, Mr Pakenham, midshipman, and myself in the other, almost despairing of being able to reach the island.

"The ship kept off and on outside the thickest part of the loose ice, through which the boats were obliged to be very careful in picking their way, on the S.E. side, where I thought I might have ascended. We reached the island, and found running on it a very heavy sea; the first lieutenant, however, landed, having backed his boat in until he could get foothold (without swimming), and then jumped overboard. I followed his example; the others were anxious to do the same, but the sea was so high that I could not permit them.

"We hoisted the jack and took possession of the island with the usual ceremonies, in the name of Her Most Gracious Majesty Queen Victoria.

"The extent we had to walk over was not more than thirty feet. From this space, and a short distance that we scrambled up, we collected eight species of plants; specimens of the rock were also brought away.

"With the time we could spare and our materials, the island was perfectly inaccessible to us. This was a great disappointment to us, as from its summit, which is elevated above the sea 1400 feet, much could have been seen, and all doubt set aside, more particularly as I knew the moment I got on board I should be obliged to carry sail to get off the pack, and out of the bight of it we were in; neither could I expect that at this late period of the season the weather would improve.

"The island on which I landed is four miles and a half in extent east and west, and about two and a half north and south, in the shape of a triangle, the western end being its apex. It is almost inaccessible on all sides, and a solid mass of granite. Innumerable black and white divers (common to this sea) here found a safe place to deposit their eggs and bring up their young; not a walrus or seal was seen on its shore, or on the ice in its vicinity. We observed here none of the small land-birds that were so numerous about us before making the land.

"It becomes a nervous thing to report a discovery of land in these regions without actually landing on it, after the unfortunate mistake to the southward; but as far as a man can be certain, who has 130 pairs of eyes to assist him, and all agreeing, I am certain we have discovered an extensive land. I think, also, it is more than probable that these peaks we saw are a continuation of the range of mountains seen by the natives off Cape Jakan (coast of Asia), mentioned by Baron Wrangel in his 'Polar Voyages.' I returned to the ship at 7 P.M., and very reluctantly made all the sail we could carry from this interesting neighbourhood, to the south-east, the wind at the time allowing me to lie just clear of the pack."*

* An American report of a later date denies the existence of this extensive land, of whose existence Captain Kellett says he feels pretty certain; but until some one actually sails over the spot, we have as much reason to believe those who saw land as those who did not.

This land some geographers suppose to extend in continuous or broken masses to the east and northward, and to form a portion of the vast archipelago of islands lying north of Barrow Strait; and directly that Captain Kellett's discovery was reported in England, it gave rise to a strong hope that Captain Collinson's expedition would be able to reach it, and follow along the southern coast towards Melville Island, as Sir Edward Parry had succeeded in doing from the opposite direction, thirty years before.

CHAPTER V.

THE INVESTIGATOR GIVES UP THE HOPE OF MEETING THE ENTERPRISE, AND DEPARTS ALONE—THE FIRST ICE—IMMENSE HERD OF WALRUSES—MOTHERS AND BABIES—VALUE OF THE WALRUS TO THE ESQUIMAUX—A BLIND LEAD—CAPE BARROW DOUBLED—PROCEEDING IN A NORTH-EASTERLY DIRECTION—GREAT EXCITEMENT—THREE ESQUIMAUX MET WITH—THEIR ASTONISHMENT AT SIGHT OF THE VESSEL.

THE most extensive body of "open water" in 1850 lay in the direction of Point Barrow, the turning point of America, a position the Investigators were impatient to reach. The Enterprise had not yet been seen; but the fogs had been dense, and the weather unfavourable for meeting her, so that all conspired to make the anxious men and officers conjecture that she was still far ahead, and waiting for them. Captain Kellett, the senior officer, was not so sanguine as to the Enterprise having made an equally quick passage; yet he felt the responsibility he should incur, should she have passed, by detaining her consort. Captain M'Clure, too, pointed out how valuable every hour was to him, and to the important service he was upon; for he well knew the force of the arctic maxim, that a day lost often entails a whole season of fruitless labour. At last Captain Kellett consented that the Investigator should part company; but he first supplied Captain M'Clure's wants, by giving him three volunteers, and furnishing him with such articles as his own stores would admit of. The reader will sympathise with the generous feelings of those who, like the captain and officers of the Herald, were thus for the last time, perhaps, in this world, shaking by the hand men bound upon a service as hazardous as it was glorious, and they will understand how trying a moment it must have been for one circumstanced as Captain Kellett was, to say to such a body as the Investigators—"Go on!" when he knew full well that from where they then stood there lay before them, for 900 miles, upon the one hand a shoal and dangerous coast, upon the other a heavy and hopeless sea of ice.

The Investigator had not long borne up on her solitary course under a heavy press of sail, when the signal was made,—"Had you better not wait forty-eight hours?"

The reply was characteristic: "*Important duty. Cannot, upon my own responsibility.*" In a few hours the Investigator was alone, and pressing on into the polar sea.

On the 2d August, in the morning, the first ice was sighted extending across ahead, in latitude 72° N. On reaching the ice immense herds of walruses were seen basking upon the loose masses: huge bulls, with splendid tusks, which would have delighted the eye of a Gordon Cumming; females, with their numerous cubs playing about, formed a sight novel and interesting even to the old Greenland ice-master. A gun was at first loaded with grape and canister for the purpose of shooting some of them; but the order was countermanded by Captain

M'Clure, from the kindly feelings awakened by the affection evinced between the mothers and babes of this brute community. Some of these creatures were conjectured to weigh as much as thirty-five hundredweight; and the ice when relieved of their weight rose about two feet.

These ferocious-looking creatures are found in great numbers in Behring Strait; all our voyagers speak of them; and the well-known sketch in Cook's Voyages of the conflict between his men and the walruses, has been seen by most people. It is only fair to observe, however, that this representation does some injustice to a brute whose character is naturally inoffensive, although, when assailed in the water, it is not deficient in courage. If in company with the female or its young, the self-devotion evinced by the bulls excels that of most animals. Both male and female have tusks; but these are so situated as to be of but little use when the creature is out of the water, unless for aiding them in scaling the steep and rugged sides of ice-hummocks. The females are sometimes seen with two cubs at a time, but more generally with only one. They suckle their young; and, from the different sizes and periods at which they have been seen doing so, voyagers are led to believe that, for twelve or eighteen months, the young one is dependent upon the mother for nourishment. They feed upon the submarine plants of the arctic regions; and as far as may be judged from the teeth, they do not appear to eat fish or seal, although it is sometimes asserted that they do so. Their thick skins, plentiful blubber, wholesome flesh, and ivory tusks, render the walrus a valuable prize to the Esquimaux wherever they are found; and in

Behring Strait a considerable traffic is carried on by the Esquimaux with the Siberian traders, in the exchange of walrus ivory for Russian knives and kettles.

Whilst we have been making this digression about walruses, the Investigator has been running on nearly due north, in the 166th meridian, or at least as much so as the ice would let her, the current aiding her the while to the extent of sixteen miles in twenty-four hours. On August 4, Captain M'Clure discovered that he was running and working up a *blind lead*, or *cul-de-sac*, out of which it was advisable to make his way by retracing his steps to the southward. This was done as rapidly as possible, and they arrived on the 5th August off Wainwright Inlet, and again sighted the Plover for a short time.

Keeping now very close to the American coast, or as much so as the ice would admit, the vessel made rapid progress towards Point Barrow. At midnight, being at the distance of about ten miles from the land, and in seventy-three fathoms water, they rounded the north-west extreme of the American continent, and began their progress to the east, and towards *home!* What joy was in that sound, and in the thought of having at any rate achieved one difficulty that had never before been mastered by a ship. On the morning of the 6th August 1850, the officers and crew of the Investigator felt free from all anxiety upon the score of being able to enter the Arctic Ocean from Behring Strait, and turn their backs upon the Pacific. That was now an accomplished fact; and a good month of the best navigable season was still before them.

Their next aspiration was to reach Melville Island; but as far as the eye could reach in that direction, a waste of ice extended, and such ice as few, if any, navigators, even in the arctic zone, had ever before seen; and the fact of a two-knot-per-hour current setting southward off Point Barrow, told Captain M'Clure pretty plainly whither the pack would drift him, if he got entangled or beset in it. To keep in the *land-water* was the only resource left; for between the American coast and the line of heavy oceanic ice, which, from its great draught of water, was checked by the shallow nature of the sea, there was a tortuous channel, varying from a few yards to a mile in width, and this land-water they hoped would lead them to that open sea off the Mackenzie River, which Sir John Richardson reported in his journey to the mouth of that river.

The wind came round to the E.S.E., with rain and mist; and against these obstacles the good ship struggled through the ice. On the one hand lay a low and dangerous coast devoid of any shelter or haven, on the other a barrier of packed ice formed of great floe-pieces and hummocks, and the intervening space was much covered with stray masses, so dense and heavy in their nature as to cause the vessel to tremble in every timber whenever she unavoidably struck any of them. Now they were sailing in a dense fog, the hand-lead and look-out man their only security against shipwreck; and anon in a gleam of sunshine and calm, towing with all their boats ahead. The excitement was great for all; and all anxiety for the future was merged in joy at the present. The men entered fully into the enthusiasm of the officers;

and loud songs and cheers rang through the solitudes of that lonely sea as each fresh difficulty was mastered, and another mile of easting attained.

On August 8, when close to Point Pitt, about 120 miles east of Point Barrow, the master, Mr Court, was sent on shore to place a notice of the Investigator having passed, and to erect a cairn. On landing, the boat was met by three Esquimaux, who, although at first extremely timid, gained confidence when the polite and pleasing operation of rubbing noses had been properly gone through in token of friendship.

Through Mr Mierching, the Moravian missionary, a communication was readily established with these Esquimaux. The three men were, it appears, a portion of a petty tribe residing in this neighbourhood. The Investigator had been first seen by them on the previous night; and as they had never seen anything like her before, she had caused no small sensation in the community, and a general meeting had taken place in consequence. The most astonishing thing of all was, how those "three great trees" (the masts) came to be moving about; and many different opinions were offered. They could give no name to this new wonder but that of "Omiack," the same they give to their large canoes. These three men had been deputed to watch and follow this wonderful Omiack; and they were the men Mr Court had fallen in with.

In reply to queries, they held out the gratifying promise of a channel of water being found continuously to the east, and that at this season it would vary from three to five miles in width. The ice, they said, never went farther off the coast than at the present time—it was then some four miles off-shore—and at one season they said there was no water at all along the coast. Yet the poor savages could give no idea of when the water ceased to exist, or when the winter season may be said to commence in this region.

Communication being established with the tribe generally, some one or two of them owned to having seen a ship before to the south (the Plover, no doubt), in Kotzebue Sound. They spoke also of trading with other natives, who gave them Russian articles in exchange for their furs, and promised, if Captain M'Clure would return, to have some skins and ivory ready for him. The tribe were a stalwart set, but the men had hideously disfigured themselves by labrets in the lower lip, most of them having two apertures on either side of the mouth, half an inch wide, into which those disgusting ornaments were thrust. The women might have been good-looking, according to the standard of thorough-bred Mongolian beauty; and some were slightly tattooed about the chin, but it was barely perceptible—any more, it must be added, than the natural colour of their faces, from an accumulation of dirt. General obesity prevailed in this arctic family, and they seemed to be in possession of stores of meat, as they offered to supply it for barter, if the ship would wait. Thieving, performed in a most artless manner, though not altogether without skill, appeared their principal accomplishment. Whilst Captain M'Clure was on one occasion serving out some tobacco as a present, he felt a hand in his trousers pocket, and on looking down found a native actually, while receiving with one hand, picking his pocket with the other. Yet, when detected, the fellow laughed so good-humouredly, and all his compatriots seemed to enjoy the joke so amazingly, that even the aggrieved parties joined in the merriment. In Esquimaux society this tendency to pocket-picking was evidently considered an amiable weakness. Captain M'Clure told these people that he was looking for a lost brother; and they promised, if they should ever find any white men in distress, to be very kind to them, and "give them deer's flesh." From some of this tribe the fact was gleaned of their having observed Lieutenant Pullen when, in 1849, he sailed along the coast to the Mackenzie River; and Captain M'Clure also satisfied himself that the Erebus or Terror had never reached this neighbourhood.

A letter was left with this tribe, to be given to any Europeans they might meet with, and they promised faithfully to deliver it, in consideration of the numerous presents they received. In proof, however, of how little faith can be placed in the promises of such savages, it was not until May 1852, when that energetic officer Captain Maguire succeeded in opening a communication with these people, that he accidentally observed a small canvass bag, directed "To the Chief Trader of the Russian Settlements, America," and eventually discovered a remnant of the original document. It only reached England in 1853, after Lieutenant Cresswell, who had been sent home by way of Baffin Bay with despatches, had reported the Investigator's safety.

CHAPTER VI.

THE NARRATIVE OF CAPTAIN MAGUIRE—THE ESQUIMAUX REPORT—THE COAST OF AMERICA IN THIS REGION—THE PACK-ICE AND FLOES—REACH JONES ISLAND—VISITED BY TWENTY-FOUR NATIVES—RECOGNITION OF THE CHIEF—SIMPLE CUNNING OF THE SAVAGES—A FAIR MALEFACTOR—GALLANTRY OF THE INVESTIGATORS—HAZARDOUS AND DIFFICULT NAVIGATION—THE DELTA OF THE MACKENZIE—THE SHIP AGROUND—SERIOUS LOSS OF PROVISIONS.

WHILE the good ship is working slowly along the coast, I may take the opportunity of showing how ably her trail was discovered by Captain Maguire in 1852-3, and by what a faint clue he ascertained the fact of both the Investigator and the Enterprise having in successive years been seen by these Esquimaux.

Captain Maguire, who succeeded the former captain of the Plover depot-ship, by dint of much energy and perseverance, forced her, in the year 1852, round Point Barrow, and reached the haunts of the natives visited by Captain M'Clure in 1850.

Captain Maguire's admirable narrative, which I have given in the Appendix, is one of the most interesting that has been written by any arctic navigator: it describes his arrival amongst the Esquimaux, and his preparations for winter; their thievish propensities; his difficulties and noble forbearance. "One day," to use his own words, "returning across the bay to the ship, we were accompanied by a young man and a boy, who talked a great deal more than we could understand; but the former, in explaining to us the sort of tobacco that had been given him on board a ship, *twisted his fingers together so as to describe American twist or negrohead.* This led us to believe that the vessel where he had obtained it might have been the Investigator or Enterprise.* . . . The two men willingly accompanied us on board; and I was glad to avail myself of Lieutenant Vernon's knowledge of the language to sift the story more thoroughly. He allowed them most patiently to describe all they had seen, in their own way, and eventually ascertained that the ship they had been on board of had diagonal † decks, and had an ice-choke ‡ larger than the Plover's. The illuminators in the deck, they remarked, were square. These are the points that seem to have caught their attention; and these were sufficient to show that they had been on board one or other of the ships; but when the captain was described

* Ships generally carry tobacco in the *leaf* for the use of their crews; but in 1850 manufactured tobacco was adopted in arctic discovery ships, to economise stowage; and *cavendish,* or flat compressed slabs, and *negrohead,* or twisted sticks, were supplied by the Admiralty to Captains Collinson's and Austin's expeditions. A little clue like this enabled Maguire to decide whether the reports he was gleaning referred to the expeditions of Franklin or Collinson.

† The planking of ships' decks is generally in lines parallel to their keels; but in arctic ships the doubling is placed diagonally across the original deck, to give a greater degree of strength in the event of being nipped by the ice.

‡ A strengthening piece of wood which goes round the ship outside.

as wearing spectacles, Captain Collinson was at once identified.

"It is worthy of notice," remarks Captain Maguire, "that a particular kind of tobacco, with which we knew the Enterprise and the Investigator to have been provided, led to a voluntary description from the Esquimaux of their having boarded a ship, thus affording more information in a few minutes than all our inquiries of the chiefs and others in several months had done;" and, we may add, had not the Resolute and Intrepid, by great good fortune, been sent to Melville Island in 1852 *via* Greenland, we should, for our information concerning the success of the Investigator in reaching as far as Point Pitt, have been mainly indebted to the keenness and skill of Captain Maguire, and the observation of a savage upon the peculiarity of her tobacco.

To return to the voyage of the Investigator. The 9th of August was passed in working against an adverse wind, through very narrow and intricate lanes of water, the ship seldom more than five minutes upon one tack, and so close to the land as to allow the natives to be constantly visiting her. Happily it insured a close and careful search of the coast for any signs of Franklin's crews having passed.

The coast of America in this neighbourhood is described as one vast plain, the soil a dark blue clay, without a stone or elevation to break its strange monotony. From the beach the eye ranged over an immense green flat, variegated with moss, grass, and flowers, and broken here and there by fine sheets of fresh water. Large herds of reindeer were seen from the Investigator.

They were a sight strangely novel to our navigators, and totally unexpected in the near neighbourhood of a sea of eternal ice. The bottom of the sea partook of the level nature of the land; and the soundings were very regular, enabling the Investigator to work along in spite of fogs, by alternately standing into three fathoms water, and off to six or seven, where they generally found the edge of the heavier ice brought up and aground. The enormous thickness of this oceanic ice may be best appreciated by remembering that to be aground in seven fathoms water the floes must have ranged from *thirty-five* to *forty* feet in depth; and this ice, from being the outer edge of the pack, was, of course, lighter than the rest. In vain, over this vast wild scene of ice, did the aching eyes of the Investigator's crew look for one glimmer of a water sky. The natives, whenever they were questioned by Mr Mierching, told the same tale. They knew of no lands north of them, and it was not possible in their seal-skin kyacks to go far into the pack. Sometimes they had been up lanes of water, which formed in the ice, to the north; but they had never been able to advance more than a day and a half's journey (or thirty miles), and this only under the most favourable circumstances. They then came to ice which forbade all further progress. There were in that part, they said, no seals to be seen; and seals would be their only inducement for such perilous voyages. This great ice, which the Investigator had afterwards to battle with, appalled even a race whose lives were spent in its neighbourhood.

As they approached the vicinity of the Colville River, its influence became plainly perceptible, in rendering the water brackish and muddy; and water-fowl, such as the common and king eider duck, were very numerous, especially in Harrison Bay.

Hitherto the current, since rounding Point Barrow, was found to be at the rate of eighteen miles a-day in an east-by-north direction; and it, as well as the smoothness of the narrow lane of water, enabled the Investigator to work to windward at the rate of thirty or forty miles a-day—a rapid progress for her. Every precaution that the invention or ingenuity of individuals could devise was adopted to insure that the fact of the ship having advanced so far should be made known to those who might hereafter communicate with the natives in the neighbourhood. The name of the ship was stamped upon the knives, and cut with a diamond upon the looking-glasses that were given away; and Captain M'Clure, by way of preventing the people from obliterating the name from the knife-blades by sharpening or polishing them, told them that the letters there engraved would be a charm to make the hunter fortunate so long as they remained intact—one of Dr Paley's justifiable fibs.

On the 11th of August, the temperature of the air and sea-water, which had been for some days strikingly equable, ranged about 34° Fahr. in the shade, and the sea 33° Fahr. This date may be regarded, therefore, as the height of summer on this coast. On Jones Island, in lat. 70° 33′ N., and long. 150° 16′ W. of Greenwich, the officer of the watch reported a spar erected as if for a signal; and considerable anxiety was felt, until Mr Court went to examine it, and reported it to be merely a piece of drift-wood forced into that position by the pressure of the ice. The beaches here were found strewed with drift-wood; and one spar was observed which was as large as the Investigator's mainmast. The centre of Jones Island was one great swamp, the breeding-place of large flocks of wild-fowl.

They were here visited by two baidars, containing twenty-four natives; and it is remarkable that, from the chief having in his possession an old musket with the date "1840, Barnet," marked on the lock, we are able now to trace the fact that his tribe hunt over the ground from Point Barrow thus far to the eastward: for it was this very man with whom Captain Maguire had such difficulty, as he mentions in his narrative (given in the Appendix); and there is little doubt that it was he who headed the attack on Lieutenant Pullen in 1849. However, he was civil enough now; and both he and his compatriots made a most favourable impression on Captain M'Clure. "The size of the vessel," he says, "and particularly that of the *large handkerchiefs* (the sails), excited their admiration. The whale-boats, as coming more within the grasp of their conception, were much admired; and they expressed curiosity to know where trees fit to make such boats out of grew, believing, as they did, that they were in one piece, and merely hollowed out." They readily parted with whatever fish and wild-fowl they had, in barter for tobacco, and offered, if Captain M'Clure would stay, to bring abundance of venison, which they had in store upon the main.

The currency of this region is tobacco; and Captain M'Clure became his own Master of the Mint, by cutting the sticks into pieces about three inches long, and paying with them as he thought just. An amusing instance was noticed on this occasion of the cunning of these savages. One of them having observed that every fish was paid for by one piece of tobacco, with a view of increasing his profits cut his fish into pieces, and with barefaced assurance proposed that he should be paid the same amount of tobacco for each portion as would have been paid for the whole. The trick afforded great amusement amongst the crew of the Investigator; and the fellow himself seemed to enter into the joke when he found his *ruse* unsuccessful.

Another party of natives, in the course of the same day, succeeded in obtaining an interview with the Investigator by hanging out the rather original ensign of a pair of seal-skin trousers from the top of a pole. After exchanging the usual signs of peace, by holding up hands on either side, and by a general chorus of Timouh ! or Peace, the more affectionate salutation of rubbing noses was gone through, and cheerfully too, for these Esquimaux, wonderful to state, happened to be this time pleasantly clean. Through the aid of Mr Mierching, the natives informed the Investigators that they had never before seen a European ; and they promised to take care of any who should ever come that way. These people barter their furs with the natives westward of them for Russian products ; but, remote as they are, few articles ever reach them, for even a knife was an object of the greatest possible delight and wonder. Their summer residence and hunting-grounds were on the sterile islands lying off the coast ; their winter lodges were a short distance inland upon the main.

The tendency to theft was the prevailing vice among these Esquimaux, as is naturally to be expected from all savages who for the first time see such incalculable wealth, in the shape of wood and iron, thrown almost into their possession; and the same may be said of all the tribes the Investigator fell in with along this coast. Even after receiving the most lavish supply of presents, and when apparently unable to express all their delight and gratitude, the temptation of any loose article was too great for their honesty. Nothing ever was "too hot or too heavy" for them. On the 12th, for instance, whilst every kindness was being shown to a party of men and women, one of the ladies contrived to secrete under her ample proportions a couple of iron winch-handles and a small ice-anchor. She settled over them like a hen over a nest of eggs; but the ends of one of the handles peeped out, and a lynx-eyed corporal of marines detected what would in those regions have been a serious loss. The fair delinquent, when taxed with the offence, and upbraided in that rough but energetic language which seamen believe must be understood from pole to pole, pointed at her husband, and evinced a wish that he should share in the responsibility. Of course gallantry was at a high pitch amongst Europeans who had left all womankind behind them so many months since, and might shortly again do so for a length of time painful to contemplate; and it is amusing, in reading over the journals of the officers, on the same page with the above anecdote to find such a passage as the following:—"These children of nature, inhabitants of one of her most desolate regions, appeared free from vice; and evinced the liveliest marks of gratitude for the trifling presents we made them!" How much such amiable forbearance, even in the passing of judgment upon these creatures, tells us of the tender-heartedness of this body of resolute men! They found a pleasure even in communicating with the veriest savages on earth as a relief to the monotony of the voyage; and probably the anticipation of success already threw its sunshine over everything they saw.

These interviews with the natives now formed the only breaks to the daily routine, beyond the general satisfaction felt when, every day at noon, the able master, Mr Stephen Court, reported so many more miles achieved to the eastward towards the Atlantic Ocean.

On August 14, the Investigator had reached longitude 148° 17′ W., and became much hampered amongst the low and, for a ship in thick weather, dangerous islands which line the coast in this neighbourhood. They had now passed the Return Reef of Sir John Franklin, the spot from whence he bore up, in 1826, after his unsuccessful attempt to reach Behring Strait from the Mackenzie River; and the Investigator might be said to be nearing the delta of that great stream.

The navigation, always most anxious work for the responsible officers, now became, if possible, still more so. Hazardous shoals were in some places hidden by floes which had run over them; in others, the soundings altered so abruptly as to deceive the most careful. Sudden and dense fogs, with changeable and sometimes rapid currents, all tended to give the commander and master every cause to be watchful. The shoals lining this American tundra are of the same character, Captain M'Clure assures us, as those at the mouths of many large rivers in various parts of the world, but which are especially remarked by Baron Wrangel as lying off the mouths of the great Asiatic streams that debouch into the Arctic Sea. They are composed purely of drift-wood and the alluvial deposits of neighbouring rivers. A mass of the former takes the ground, or becomes fixed by some accident, in three or four fathoms water; the current soon feels the impediment, and begins to deposit, in and around the nucleus, matter that forms a shoal; the shoal grows rapidly, more drift-wood grounds, more sediment is deposited, and even within the lifetime of a man, as one Esquimaux assured Mr Mierching, an island rises from the bottom of the sea.

After one or two narrow escapes on the 14th of August, the Investigator found herself quite beset with these shoals; and at last, in trying to escape through a narrow three-fathom channel, she unfortunately took the ground; a press of canvass was at first carried, in the hope of dragging her through it, but that proved to be hopeless; a kedge anchor was next laid out, but it, as well as a stream one and chain, failed to heave the vessel afloat. No time was now to be lost; the boats were got out, all the deck-load of provisions embarked in them, some tons of fresh water were started from the tanks in her hold, the anchors lowered into cutters, and then, happily, on a fresh attempt being made, the Investigator was floated, after being aground for five hours. The ship was, however, obliged to carry canvass to get into deep water, and, in doing so, one of the boats laden with provisions capsized whilst in tow, and eleven casks of salt meat were lost,—a real calamity to people situated as they were.

CHAPTER VII.

A THUNDER-STORM — SLOW PROGRESS—SIGNS OF THE NEAR APPROACH OF WINTER—GOING AHEAD AGAIN—THE SHIP RUNS INTO A TRAP IN THE MAIN PACK — WORKS OUT AGAIN — COMMANDER PULLEN AND A BOAT'S CREW PASS WITHOUT BEING SEEN—LAND AT POINT WARREN—HOSTILE RECEPTION —RECONCILIATION.

THE 15th of August was a lost day; and, entangled by shoals of ice, the vessel had to anchor off Yarborough Inlet, to avoid again getting ashore. In the evening came a westerly wind, and with it a hope of release, for the drift-ice had shut up all the navigable channels. This wind was attended by a thunder-storm, a phenomenon rarely witnessed by an arctic navigator. "The west wind," writes Captain M'Clure, "was ushered in with rain, and *thunder* and *lightning*. The two latter I never before witnessed in such a latitude (70° N.) The thermometer rose to 45° Fahrenheit, and the air was quite sultry, with dark heavy clouds rising overhead, resembling those seen in a thunder-storm in our own country. The packed ice to-day," he continues, "as far as the eye can reach, appears solid and heavy, without a drop of water discernible. The refraction has been considerable, giving to the edge of the pack the appearance of a continuous line of chalk cliffs, from forty to fifty feet in height. From the light shady tint, which is distinctly visible in different parts of the pack, I should be inclined to think that there may be many of the same kind of islands as those we have met with, extending to the northward, and impeding the progress of the ice, thereby keeping this sea eternally frozen."

Scarcely any progress could be made on the 16th or 17th, though the men worked hard and incessantly,—now towing, now warping, and, when any wind served, pressing the old craft to her work until the bells in the ship rang again by the concussion of her bows against the floating masses of ice.

An attempt to catch fish with the seine upon one of the shoals, by a party of volunteers, proved abortive, nothing but drift-wood repaying them for their industry; and on the night of the 17th the surface of the sea was seen, for the first time this season, to have a coating of ice formed over it. The fact was observed and remarked upon by all: it told of the near approach of winter, with all its disagreeables, and caused some, who had limited their aspirations for this year to the Mackenzie River, to question if they should even reach so far.

On the 18th, the Investigator made some progress, and passed Flaxman Island. She was now fairly about to cross the large angle in the coast formed by the mouth of the Mackenzie River.

Tempted by what seemed a sea of water, the Investigator ran off, steering a N.E. course for Banks Land. A slight pitching motion, the first they had felt for some time, encouraged to the hope of large water. The rapid deepening of the sea, too, from nine to thirty-two fathoms, during the first watch, likewise strengthened their hopes; and darkness and a thick fog hid for a while the reality from their eyes.

August the 19th came, with a fresh westerly breeze, snow squalls, and mist; and in happy ignorance they rattled on, sighting every now and then what looked like the pack edge, or tumbling into bights of ice, where there was no way out but by returning for a while upon their footsteps. At noon they sounded in 195 fathoms without bottom; and shortly afterwards the disagreeable fact of the Investigator having run into a trap in the main pack presented itself to the mind of the captain. Ice, of stupendous thickness, and in extensive floes, some seven or eight miles in length, was seen on either hand; the surface of it not flat, such as we see it in Baffin Strait and the adjacent seas, but rugged with the accumulated snow, frost, and thaws of centuries. Ninety miles had they run into a blind lead in this dangerous ice; and if the wind should shift and the ice close, the position of the ship would be critical indeed. Captain M'Clure now hauled to the southward, and worked against the wind, which freshened and forced him at one time to carry double-reefed topsails.

Next day a distant view of the Buckland Mountains was obtained, but the ship had eventually to retrace her steps seventy miles to the southward, before she was safe from the jaws of the pack,—an escape which all were truly grateful for, there being no two opinions in the ship as to what would have been her fate had the ice closed upon the Investigator.

On the 21st August the sea was sufficiently clear within the edge of the packed ice to allow the Investigator to steer a course outside the Pelly Islands, which lie off the mouth of the Mackenzie River, and fifty miles distant from the mainland. At 10 A.M. the ship passed distinctly athwart the stream of the Mackenzie, flowing northward into the polar sea. The temperature of the sea rose from 28° to 39° Fahrenheit. The colour of it was as muddy as the Thames at Woolwich, and to the taste only slightly brackish. The depth of water was, however, but four fathoms, and warned Captain M'Clure that he must not attempt to approach the mainland any closer, anxious as he was to do so.

Little did Captain M'Clure or his gallant companions imagine that on that 22d August a European boat's crew, under Commander Pullen, was only a few miles off, on their return homeward from a visit to Cape Bathurst; but such are the unavoidable accidents of arctic service. The two following references from Captain M'Clure's and Lieutenant Pullen's journals will, by a reference to the chart, show how near they were to each other.

Captain M'Clure, on the 22d August, during the evening, observed Richard Island bearing S.E. to N.N.E. On the 23d August, in the morning, Lieutenant Pullen was steering for Richard Island, and in the course of that day he coasted along its northern shore on his way to the Mackenzie River. Can a more convincing proof be given of the difficulty of meeting in polar seas, or of such a search as that undertaken for Franklin's expedition?

For, be it remembered, both officers knew of the importance of communication between all parties employed upon such service, and certainly did their best to pass nothing unseen. Perhaps, however, it was as well for the future fame of the British navy that Captain M'Clure did not meet Lieutenant Pullen, as the latter's unfavourable report of what an ice-encumbered sea he had seen from Cape Bathurst might have induced Captain M'Clure to adopt some other course instead of the one he pursued.

After passing the Pelly Islands, whales were again seen for the first time since passing Point Barrow—the ice-master, however, said they were small, and not worth much to fishermen, either in bone or oil.

On the 24th August some native huts were seen near Point Warren on the eastern shore of the Mackenzie, and Captain M'Clure landed in the hope of inducing some of the natives to carry to the Hudson Bay Company's posts intelligence of his having passed the Mackenzie River; for our voyagers were now fairly within the limits of a region whereon, it was to be expected, the civilising influence of that wealthy Company of monopolisers ought to have been perceptible. Hitherto they had been coasting along *Russian America;* but now *British America* had been entered upon.

The reception of the Investigators by their Esquimaux fellow-subjects of Queen Victoria denoted anything but confidence in white men. Two of them yelled and shouted, waving a knife in declaration of war, and even threatening with bow and arrow. Every method which had hitherto been found of avail in propitiating the goodwill of Esquimaux was made in vain, until at last

Mr Mierching, disguised as one of themselves, succeeded in assuring them of the good intentions of their visitors, and that they neither contemplated robbing nor murdering them,—a pleasant contingency which they evidently connected with a visit from "Kabloonas," or white men. Confidence had only just been established, when suddenly a couple of muskets were espied in the hands of the boat's crew, and thereupon all the fury of the natives burst out afresh; nothing appeased them until the muskets were sent back to the boats. It appeared that, when the vessel was first seen in the morning, all the natives had decamped with their baidars and household gods, leaving only the chief and his son, who had bravely refused to fly, and remained to defend a sick youth and the encampment. This invalid soon made his appearance with his mother, and Dr Armstrong kindly sought to give the poor creature such aid as was in his power: but it was too late; his foot was evidently in an advanced state of mortification, and death must have soon come to put him out of his misery.

Through Mr Mierching they learnt that this tribe was at war with its neighbours, and had no communication with the Indians of the Mackenzie River. Their barter or trade was carried on, after the sea froze over, by crossing to the western shore, and meeting the natives seen by the Investigators on her road hither. The chief recognised the name of one of the natives Mr Mierching had previously met, and said, with some degree of pride, "Ah, he was a great chief! He should," he added, "see Attauwoo very soon," and "he had a quantity of blubber and whalebone to barter with the western people, for his people had killed three whales in the present season." When asked why they did not trade with the white men up the big river, the reply was, they had given the Indians a water which had killed a great many of them and made others foolish, and they did not want to have any of it!

From this tribe Captain M'Clure heard of two boats having been to Cape Warren from the westward, and having returned again; and he was for some time rather puzzled at a story they told him, of a white man having been killed and buried in this neighbourhood. To the inquiry of when it had taken place, all that could be learned was, that "it might have been last year, or perhaps when the narrator was a child!" an Esquimaux mode of dealing with dates not a little perplexing.

This story evidently referred to the death and burial of a man near this place, in one of Sir John Richardson's early journeys from the Mackenzie to the Coppermine River. Captain M'Clure, not having been supplied with all modern books of arctic discovery, was not aware of this circumstance.

Although the natives offered to show where this body was interred, it was not until next day that circumstances admitted of an examination being made in the locality pointed out; and then it was without any success, although the ruined remains of a couple of drift-wood huts, so well described by Sir John Richardson in his journey through Prince Rupert Land, were there to excite curiosity and afford grounds for conjecture.

CHAPTER VIII.

APPROACH TO CAPE BATHURST—WHALE-FISHING OF THE NATIVES—CELEBRATION OF THEIR VICTORIES OVER THE LEVIATHAN—ESQUIMAUX CHARMERS—THE JOYS OF CAPE BATHURST—THE LAND OF THE WHITE BEAR—AN ESQUIMAUX SWINDLER—MODE OF SETTLING QUARRELS—JUDICIOUS MISSIONARIES DESIRABLE FOR THESE PEOPLE—ADMIRABLE QUALIFICATIONS OF MR MIERCHING.

The 26th and 27th of August 1850 were spent in making the best of their way from Cape Warren to Port Dalhousie, the vessel being kept as near to the land as the soundings would admit of, about two miles. Captain M'Clure would have sent a boat to render the search more perfect; but the incessant mists and variable weather made it hardly prudent to detach any of the men, with the possibility of their being misled. The extent of open water off the land seemed to increase as they approached Cape Bathurst; but the floes that were found floating about in it were of great magnitude, and occasionally gave much trouble in keeping the ship clear of them. The nights were lengthening perceptibly; and from the perpetual day of the arctic zone in summer they had now three hours of perfect darkness, during which time guns and rockets were fired at intervals, in case any of Franklin's expedition, or the Enterprise, might be near. Arriving off the western entrance of Liverpool Bay, Captain M'Clure was very anxious to run into it, in order that he might form an idea of its fitness for winter-quarters; the necessity for which, with due forethought, he felt it right to keep in mind, the more so, that along the whole extent of the American coast he had traversed since leaving Behring Strait, not a harbour had been found. The shallow and intricate navigation of the mouth of the Mackenzie River forbade his thinking of bearing up for it at any time; and he saw full well that to winter off such an exposed coast, with the whole weight of the northern ice setting down on it during November and December gales, would be certain destruction to his ship. Liverpool Bay, however, was too shoal; and the Investigator pushed on, trusting to Providence to find a safe spot when the winter came.

Indeed, some already talked of not wintering at all; and the more sanguine pointed out that they were close upon the longitude of Melville Island, a place reached by Parry in a few days from the Atlantic in the opposite direction. That goal reached, they fancied the rest of their voyage would be easily accomplished.

Crossing Liverpool Bay, and sighting several whales, some large, but the majority small, they reached Cape Bathurst on the 31st of August. Here the depth of water near the land allowed a nearer approach; it is generally described as exhibiting, along the coast, blue clay cliffs about thirty feet perpendicular, having on their summits a good depth of rich black mould resembling bog-earth. A fine plain rolled away into the interior, rich in hyperborean plants, and abounding in reindeer, whilst, besides whales, there were seen at several places positive proofs of fish being plentiful, as well as wild-fowl. At and about Cape Bathurst, where the Esquimaux were very numerous, Captain M'Clure made a final effort to communicate his position to the Hudson Bay posts.

Aided by Mr Mierching as interpreter, and by the favourable impression which Sir John Richardson's visit to them in 1848 had made, the intercourse with this tribe, numbering three hundred souls, was extremely interesting. Even a few women who first met the Investigators showed no signs of mistrust, but cordially welcomed them, and volunteered to show the way to their companions. Captain M'Clure describes them as an extremely fine-looking body of men and women; many of the latter indeed were, according to his account, exceedingly pretty. Healthy, well fed, and well clothed, they appeared to lack nothing; and their intelligence, courage, and good-natured confidence won the goodwill of their European visitors.

The chief promised to convey a letter to a tribe that communicated with trading-posts on the "Big River" (the Mackenzie River); they bartered with a tribe, which was probably the one known as the Louchoux Indians.

As far as could be gleaned, these people would proceed south for the latter purpose in about three weeks' time, leaving only a few men and most of the women to winter at Cape Bathurst. Whaling was at present their object; and their mode of killing those leviathans was primitive enough.

An *omaiak*, or women's boat, is *manned* by ladies,

having as harpooner a chosen man of the tribe: attending on this craft are a number of *kyacks*, or single-men canoes. The harpooner singles out a fish, and drives into its flesh his weapon, to which is attached, by means of a walrus-hide thong, an inflated seal-skin. The wounded fish, hampered by the inflated skin, and forced to rise to breathe in a small area of water, is then incessantly harassed by the men in the kyacks with weapons of a similar description, a number of which, when attached to the whale, baffle its efforts to escape, and wear out its strength, until, in the course of a day, the whale dies from sheer exhaustion and loss of blood.

The harpooner, after a successful day's sport, is a very great personage, and invariably decorated with the Esquimaux order of the Blue Ribbon; that is, he has a blue line drawn across his face over the bridge of his nose. This is the highest honour known to the heroes of Cape Bathurst; but to it is attached also the happy privilege of the decorated individual being allowed to take unto himself a second wife. Great orgies occur upon such occasions; and, if all tales be true, it is to be feared that morality is at rather a low ebb in these latitudes, and that Byron's theory concerning cold climates and chastity is not always supported by fact. These "children of nature" stole, of course, when the chance offered, like their brethren farther west; and the thieves were generally of the fair sex—it appeared to be a sort of tax which they levied upon the amused, and, in some cases, admiring seamen. When they came to pay a visit to the ship, they were soon quite at their ease, and having carried up their light canoes and deposited them on deck, they ranged about full of astonishment and curiosity; the pictures and looking-glasses in the officers' cabins were especial objects of admiration. They then had a dance with the crew, and invited them ashore; and charming young ladies, with brightest of eyes and whitest of teeth, assured their admirers that all the night of the 31st of August they had been expected. Venison had been roasted, whale stewed, and other racy and tempting delicacies prepared. We are assured that some of the men, in the solitudes of Banks Land, often looked back to this oasis of Cape Bathurst with a sigh, and would have willingly exchanged, for the certainty of existence there, the then uncertain prospect of an escape to Europe. Cape Bathurst was to the Investigator, in her long voyage, what Otaheite was in the olden day to our early circumnavigators.

Great skill in delineating the outlines of the coast, or chart-drawing, was noticed amongst these people. It has been often mentioned by navigators as common amongst the Esquimaux, but still they could throw no light upon the question of what lay to the north. They did not know whether it was sea or not; but they said, pointing to it with an expression of anxiety, "That is the Land of the White Bear!" They appeared to be much alarmed, too, when the ship for a time stood off towards it. The bears they described as coming from it were said to be very fierce and dangerous; and one of the women, with tears in her eyes, told how lately one of those brutes had carried off her child when playing on the beach at a short distance from her. Even our men, fearless of bears, could not but enter into the feelings of superstitious awe with which the Esquimaux pointed at that vast and mysterious sea of ice which lay away to the north-west, a sea which ship could not sail through nor man traverse. "Rightly," says Captain M'Clure, "did they call it the Land of the White Bear."

While at Cape Bathurst a constant exchange of garments went on between the seamen and officers on the one side and the natives on the other; but one Esquimaux, more knowing than the rest, hit upon an ingenious plan to obtain clothing without giving a *quid pro quo*. He went to several individuals of the Investigator's company, commencing with the commander, and pretended to be suffering from excessive cold. His teeth chattered and his whole frame shook so, that compassion was immediately aroused, and a Guernsey frock given him; then he felt better, but, watching an opportunity, the rogue would slip it off, stow it away in his kyack, and then return to obtain a fresh one. At last, however, an old quartermaster, who had been watching him with some degree of amusement, flew into a passion at the fellow trying the same trick on with him, called him "*a Jew*," and threatened to knock his head off, accompanying his threat with a demonstration from a large horny fist, which the Esquimaux understood better than the profuse volley of adjectives that rolled out at the same time over the quartermaster's quid.

With regard to the story told of a white man being buried at Cape Warren, they merely said they knew nothing of it or of the natives residing there; indeed, they were at variance with them. They appeared to have no idea of any religious ceremonies, and knew of no Supreme Being. They were generally happy, and agreed well together in their tribe; and when any quarrel did occur, they only packed up their goods and quitted the community, settling elsewhere on the coast.

If a mortal grudge should arise, a thing of rare occurrence, the aggrieved party, concealing his passion, waited quietly for an opportunity of revenge; and when it offered, he killed his enemy. No retaliation took place at the time; but some one of the murderer's family eventually atoned for the deed: the actual perpetrator, however, often escaped. Such was the principal information gleaned from these people. A despatch was left with them, which has not yet come to hand; but they promised to be kind to any strangers, "white men," who might come amongst them; a promise that they appeared likely to keep from interested, if not from better motives.

No apology is necessary for thus relating what little is known of these interesting arctic fishermen. Cut off from civilisation by a dreary wilderness but seldom traversed—hemmed in by bloodthirsty races which all the romantic fiction of a Fenimore Cooper cannot redeem from the curse of Christian men—wandering along the farthest shores of a territory farmed to a company of furriers (the Hudson Bay Company), whose dividends depend upon the race of beasts being multiplied rather than that of men,—these poor Esquimaux deserve our best sympathy; and, judging by the progress they have made in civilisation in the Danish colonies of Greenland, they appear more likely to repay the missionary than most savages we know of.

We shall probably not hear much more of these poor

creatures now that there is a lull in arctic exploration, but we cannot take leave of them without echoing a wish continually expressed throughout Captain M'Clure's Journal:—"Would that some practically Christian body, such as the Moravian Mission, could send a few of their brethren amongst the tribes of Esquimaux who wander along the polar sea, to carry to them the arts and advantages of civilised life, and trust to God, in His own good time, showing them the way to eternal life." Such men as Mr Mierching would in a few years perfectly revolutionise such a docile and intelligent race. He was, as we have said, a native of Saxony, and had for many years been a missionary in Labrador. Nothing came amiss to this energetic man; he could make a pair of shoes, or crochet an antimacassar, build a house of mud or wood, or sing a hymn and play the organ. He was strong in frame, and cheerful and contented under all circumstances, perhaps partly because he had always been accustomed to a life of trial. Such a man as he would be worth a hundred of the pretenders to piety who have fallen, like locusts, on the loaves and fishes of many of the races of uncivilised man, under the plea of "plucking brands from the burning."

CHAPTER IX.

CAPE BATHURST LEFT—FIRES OBSERVED ON SHORE—PROVE TO BE VOLCANOES—CAPE PARRY REACHED—NEW LAND DISCOVERED—POSSESSION TAKEN IN THE QUEEN'S NAME—THE INVESTIGATOR PROCEEDS ON A NORTH-EASTERLY COURSE—BARROW STRAIT ONLY SIXTY MILES FARTHER — CAPTAIN M'CLURE'S JOURNAL.

The month of September found the Investigator still pushing ahead, but much delayed by the light winds which had prevailed ever since passing Behring Strait, except for the few hours the ship had been entangled in the pack off the Mackenzie River.

From the 1st to the 5th the vessel was coasting round the bay formed by Capes Bathurst and Parry; whales were very numerous, no less than fifteen being seen at one time, although none of a large size. The water was deep, eighty-four fathoms being obtained, only four miles off shore, when near the mouth of the Horton River. On the 4th large fires were seen on shore, and at first supposed to be lighted by the natives to attract attention. Mr Mierching, however, questioned such extravagance in fuel being committed by Esquimaux, and on the next day (Sept. 5), when it happened to be calm with rain, Lieutenant Gurney Cresswell, Doctor Armstrong, and some others, were sent to examine the spot.

The general appearance of the land was flat, though rising in places to an elevation of 300 to 500 feet, and intersected with ravines exhibiting blue clay. The fires proved to be volcanic, and issued in smoke strongly impregnated with sulphur, from fifteen different cone-like apertures resembling lime-kilns, whence Dr Armstrong collected a considerable quantity of specimens of earths and minerals, in which the locality was rich. The volcanoes were about fifty feet above water, and situated on an old landslip, not unlike the undercliff of the Isle of Wight: some pools of water near these volcanic cones were strongly impregnated with copperas; and the testimony of our voyagers would lead us to suppose that these subterranean fires have a different origin to those found existing here and there, in about the 56th parallel of north latitude, on the western side of the Rocky Mountains, where such fires are generally imputed to a substratum of coal having caught fire by spontaneous combustion.

On 6th September Cape Parry was reached while a fresh breeze was blowing, but the pack was still seen extending east and west in a close and heavy body, about three miles off shore, barring the road towards the Parry Group, whither M'Clure now sought to press on. At noon the sky lifted a little to the northward, and showed high bold land lying off to north-east, the extremes of its bearing N. by E. and E.N.E. true. It was an important discovery, for hitherto our chart had been a blank in that quarter. It was satisfactory, too, on other grounds, for the new land was upon the same bearing as Melville Island, and Captain M'Clure knew full well the advantage and the prospect of reaching the latter, if this discovered land was an extensive one.

Land-water had already brought him nearly half-way to Baffin Bay; next season, if not in this, land-water would enable him to achieve the rest, he naturally argued.

A freshening north-east breeze and clear weather, with open water, enabled the gallant Investigator to stretch off from the American continent during this night; and the water became more free from ice as they reached under the weather and newly-discovered land; and next day, 7th September, at 9.30 A.M., Captain M'Clure landed to take possession of this addition to the realms of his Royal Mistress. He has been blamed for this; but few men would have hesitated to do the same under similar circumstances. The devotion and enterprise which had brought that company of sixty British seamen so far, were alloyed by no selfish vanity; there were none but themselves in those wild solitudes to re-echo their cheers; and the loyalty with which they hailed their first addition to Queen Victoria's broad realms, was as sincere as that which had buoyed them up in past difficulties, and cheered and invigorated them on for future trials. It was not for them to weigh the value of what they had discovered, it was enough that they had done their duty; and an honest conviction of that fact must have gladdened officer and man that day as they stood on the cape which marked the half-way of their journey. They felt that, although they might not be the men fated to tell their own tale, and to reap the reward of their toils; yet, come what might, they trusted that if at some future day their country should learn how honestly they had devoted their lives to her glory, she would not fail to do honour to their memory. Such were, we feel sure, the high and ennobling thoughts which filled the hearts of the majority of that little band: well might their leader feel proud of them, and they of him; and both can well afford to smile at any attempt to rob them of their honest fame, or sneer at their just enthusiasm.

They christened the land "Baring Island," after the then First Lord of the Admiralty, Sir Francis Baring, under the supposition, afterwards found erroneous, that it was not connected with Banks Land. The headland or cape they were standing upon is a remarkably striking one, full 1000 feet high, and of a castellated appearance; it was appropriately named after Lord Nelson, who, as a dead hero, has not been sufficiently remembered by modern naval discoverers.

Blest with a southern aspect, the vegetation of Baring Land, for so high a latitude, was somewhat abundant, and the arctic flora was seen in perfection. Recent traces of reindeer and hares were a satisfactory sight to the Investigators; and some wild geese were soaring over head. The ice, which was beginning to make in the pools and on the land, had sent the ducks to milder regions southward; but that they came here in large numbers during the summer months, was very evident.

Better than all, too,—for who contemplated wintering there?—from a considerable elevation which, our navigators guessed, embraced forty miles of horizon to the north-east, the sea was perfectly open and free from ice. Oh for a fair wind! was the exclamation that burst from all lips.

The vessel was now worked along to the north-east, against a moderate east wind, the weather alternately foggy and fine. The coast of Baring Land, as they advanced, showed out point after point, the outline generally picturesque, and sloping to the sea. Limestone prevailed, but was covered with a considerable amount of verdure. Throughout the 8th September the Investigator advanced as rapidly as her slow sailing qualities would admit of; the soundings varied steadily from nine fathoms to seventy-six fathoms, the bottom a dark mud, and in places yellow clay. The marked continuance of the land began now to lead them to suppose that its connection with Banks Land was possible; and when next day, the 9th, after a shift of wind had enabled the Investigator to run awhile upon a straight course to the north-east, and land showed out on the starboard bow, great anxiety was felt by some lest they should be running up into some deep fiord or inlet without an egress into Barrow Strait.

Should this prove to be the case, they would have to retrace their steps; and as the season for navigation was now to be told in hours, the anxiety of all on board was for awhile intense. Hour after hour, however, passed: there was still a channel ahead! Doubt was tossed to the winds, and, mutually congratulating one another, the officers declared the land on the starboard bow to be another island, and named it Prince Albert Land. The distance between these two newly-found lands was conjectured to be thirty miles, and at an equal distance from each the Investigator held on her lonely way, in spite of fogs and snow-squalls. A few gulls and seals were seen, and some ducks flying south—an unerring signal of the advent of an arctic winter.

"The soundings in midchannel were about thirty-seven fathoms, mud," says Captain M'Clure; "and on the evening of the 9th no snow was yet to be seen lying on the adjacent land. On September 10, in a fog, the ship fell in with two islands, and it was afterwards seen that the strait they were going up contracted here to only fourteen miles, and some ice was seen hanging about the western shore."

"September 9, 1850.—Albert Land, on the starboard hand, exhibited, in its interior, ranges of mountains covered with snow; but the lower grounds were as yet free; here and there peaks of a volcanic character and outline were seen, but none that appeared active, and the rocks of the coast-line were mostly limestone, as on the western shore." Among the islands named after H.R.H. the Princess-Royal, sea-gulls still lingered, giving a hope of winter having delayed its arrival, for winter was now what the voyagers most feared. They felt as if they would give all they held dear in life for another week of summer. The dangers of the navigation, the risk of hunger and hardship—all were forgotten. "Only give us time," they said, "and we must make the north-west passage." At noon the observations placed the Investigator only *sixty* miles from Barrow Strait. "I cannot," writes Captain M'Clure, in his private journal, "describe my anxious feelings. Can it be possible that this water communicates with Barrow Strait, and shall prove to be the long-sought north-west passage? Can it be that so humble a creature as I am will be permitted to perform what has baffled the talented and wise for hundreds of years? But all praise be ascribed unto Him who hath conducted us so far in safety! His ways are not our ways, or the means that He uses to accomplish His ends within our comprehension. The wisdom of the world is foolishness with Him."

Captain M'Clure, I am sure, need be under no apprehension that his feelings, and those of his gallant supporters, will not be properly appreciated. One such paragraph as that above quoted is enough to show how well in that hour of joy, as well as in future ones of anxiety and distress, they both placed their trust where there could be no disappointment. An eloquent tribute to this truly chivalrous dependence upon God and a good cause, has been furnished by a Continental writer.* He says, after quoting such a paragraph as the above: "Le sentiment intime de la Bible si commun aux Anglais, les suit partout; il les accompagne dans toutes les épreuves, les soutient dans tous les dangers. Quand le Calife Omar brûla la bibliothèque d'Alexandrie, il dit: 'Si les livres ne contiennent que le Coran, ils sont inutiles; s'ils contiennent autre chose, ils sont de trop sur la terre.' Ainsi les Anglais avec leur Bible, ce livre unique leur suffit: il contient tout. Et quand on les suit dans ces courses héroïques qu'ils font dans les régions inexplorées, on ne peut s'empêcher d'ouvrir avec eux le Livre des livres. Ces intrépides pionniers, ces précurseurs de la civilisation qui ouvrent à l'humanité de nouvelles voies, nous apparaissent comme des Moïses qui vont à la conquête de la terre promise."

* M. Lemoine in 'L'Indépendance Belge.'

CHAPTER X.

SIGNS OF A RAPIDLY-APPROACHING WINTER—CRITICAL POSITION OF THE INVESTIGATOR—MADE FAST TO A FLOE—SAFE FOR THE PRESENT—WINTER BEGUN—WINTER CLOTHING—DRIVEN WITH THE ICE TOWARDS BARROW STRAIT—ARDUOUS TOILS—THE INVESTIGATOR REACHES HER MOST ADVANCED POSITION—BESET AT LAST—DANGEROUS AGITATION IN THE ICE—PREPARATIONS FOR SHIPWRECK—SWEEPING WITH THE PACK AGAINST THE CLIFFS—IMMINENT PERIL—SAFE ONCE MORE—AND STATIONARY.

THE 11th September 1850 came in upon the Investigator, and brought with it the first undoubted signs of winter. The ice, acted upon by a fresh north-west gale, rolled down the strait, and beset the ship, its motion being at times appallingly rapid. The thermometer fell to 21°, or 11° below freezing-point; and long dark nights added to the difficulties of navigating in such inclement weather. Harbour or winter-quarters fit to secure the ship in there were none in sight; and if there had been, it would have been out of the question as yet to retreat upon one whilst Barrow Strait was so near at hand.

On that very same day, Captain Austin's* expedition, which it will be remembered left England shortly after Captain M'Clure's did, to reach Melville Island from Baffin

* See 'Stray Leaves from an Arctic Journal.' By the Editor.

Bay, was overtaken by similar signs of winter off Griffith Island, the position of the two parties (each ignorant of the other's whereabouts) being about 400 miles from each other in an E.N.E. and W.S.W. direction. Strangely enough, too, in proof of the fact that the seasons in the frigid zone agree very much year after year, it was exactly two years anterior, upon that very day, that the expedition of Sir James Ross was frozen in permanently in Leopold Harbour.

The Investigator's position was now most critical; for the westerly gales had caught her upon the eastern and lee shore of Prince of Wales Strait, and pressed her, together with the ice with which she was surrounded, down upon that coast. Her only safeguard from destruction, for some time, was in holding on, with strong hawsers and stream chain, to ice-anchors fixed in a heavy floe, which, from drawing more water than the ship, served, when it grounded, as a natural breakwater for her against further ice-pressure.

Along the westward side of the strait, the gale caused a fine lane of clear water to be formed,—a tantalising sight for the imprisoned officers and men! It served, however, to feed anticipation, and to prevent their leader from thinking of winter-quarters.

On the 12th September his journal is to the following effect:—"The temperature of the water has fallen to 28° Fahr. (freezing point of sea-water). The breeze has freshened to a gale, bringing with it snow, and sending down large masses of ice upon us. The pressure is considerable, listing the vessel several degrees. Fortunately a large floe, which was fast approaching the vessel, has had its progress arrested by one extreme of it taking the ground and the other end locking with a grounded floe upon our weather beam. It is thus completely checked, and forms a safe barrier against all further pressure. As the rudder was likely to become damaged, it was unhung, and suspended over the stern. We can now do nothing, being regularly beset, but await any favourable change of the ice, which we anxiously look for, knowing that the navigable season for this year has almost reached its utmost limit, and that a few hours of clear water will, in all probability, solve the long-sought problem as to the practicability of a north-west passage."

The 13th and 14th September brought no change for the better; the ice, acted upon by winds, tides, and currents, kept in constant motion outside of the Investigator, and gave rise to illusory prospects of open water and fair leads. By dint of great labour and watching for favourable opportunities, the ship was gradually warped, and hauled about 1200 yards farther off shore and to windward.

The temperature of the air fell to 10° of Fahr., or 22° below freezing-point; the surface of the sea, where free from pack or broken ice, congealed and froze rapidly; the land became hidden under a general covering of snow; the stern reign of an arctic winter had begun.

Winter raiment was now generally adopted; and more than one anxious wish was expressed for some sheltered cove to heave in sight, wherein the risk of being drifted with the pack of Prince of Wales Strait during the long night of a polar winter might be avoided.

The likelihood of such an occurrence forced itself disagreeably upon the minds of all who looked in the direction of the Princess-Royal group, and saw those dark cliffs ripping up the ice which rolled down upon them. For the veriest arctic tyro knew, enchained as the ship was in the pack, if she should touch the ground before the adjacent moving body of ice did, it would roll over her, entailing certain destruction of the ship, and at such a rigorous season most assuredly a great loss of life.

The appearance of a few of the hardier gulls of those regions cheered the men a little; and the captain remarks that the appearance of these birds was a good omen, which he believed to indicate water somewhere near him; and yet he does not deny that every day now lost through the ship being beset in the ice added to his intense anxiety. Should he be forced to winter in the pack, and escape shipwreck, he could not tell where he might be drifted to in the coming winter.

On the 15th September, the wind veered a little more to the southward, or up the channel, and the ice began to drive towards Barrow Strait, and opened a little at the same time. All hands zealously set to work to reach the largest spaces of water in sight: and this labour was pursued even during the night, the men in the dark leaping and carrying the hawsers from piece to piece of ice, trusting to its white glimmer to find their road and secure a footing. Drifting along in a churning sea of ice, amid darkness and snow-storm, the Investigator held her way, her gallant company contented to run all risks, so long as her course was onward, and towards the north-east; yet the sudden variations of the soundings which the men in the chains called out, sometimes only as little as five fathoms water, and then off again to twenty, reminded them of how perilous was the course they were pursuing.

On the 16th September, the ship was still making slow progress towards Barrow Strait; but on the 17th September 1850, H.M.S. Investigator reached her most advanced position in lat. 73° 10′ N., and long. 117° 10′ W., about *thirty miles* from the waters of that series of straits which, under the names of Melville, Barrow, and Lancaster, communicate with Baffin Bay. At this tantalising distance, the ship ceased to drift, and the ice appeared to have reached a point beyond which, from some unknown cause, it could no longer find a vent. The heavy pack of Melville Strait lying across the head of the channel, was supposed to be the reason of the ice of Prince of Wales Strait thus ceasing to move on to the north-east; and the impassable nature of that heavy pack in the following year, confirmed this hypothesis. On the 19th September 1850, Captain M'Clure tells us he debated in his mind whether to abandon all hope of reaching Barrow Strait that year, and retrace his steps southward in search of a wintering-place, or to hold on, so far as he might, and run the risk of wintering in the pack. "I decided," he says, "upon the latter of these two courses;" and the consideration which influenced him in so difficult a choice was, "that to relinquish the ground obtained through so much labour and anxiety, for the remote chance of finding safe winter-quarters, would be injudicious, thoroughly impressed as I was with the absolute importance of retaining every mile, to insure any favourable results while navigating these seas."

Besides this, it was desirable to hold as advanced a position as possible, in order that the spring sledge-parties which he contemplated despatching in 1851, should at once set to work upon new and unsearched coast-lines. To winter voluntarily in the pack was, therefore, as resolutely decided upon as if arctic authorities had never said that such an attempt would result in certain destruction; and that same reliance upon an overruling Providence, which had carried them successfully so far, cheered them still in the anxieties to which their novel experiment was about to give rise.

The smallest pools of water in the strait now became rapidly covered with new-formed ice; the eider ducks, the hardiest and strongest-winged of the feathered tribes visiting the polar seas, were last seen on the 23d September; and the temperature of the air fast verged towards the zero of Fahrenheit. Although the sea-ice had formed round the ship, and the pack was re-cemented to a certain degree, still it was far from quiescent. Sometimes a pressure would take place upon opposite sides of the body,—which was still detached from the coast of Banks Land as well as the eastern shore,—the sheets of young ice would crack across, and one part overrun the other with a sharp chirping noise, which reverberated through the frosty air; at another time some huge field of ice, which from its great depth or thickness was more acted upon by the tides or currents than its neighbours, would rush with fearful velocity through the lighter ice, turning up everything that came in its way, and giving rise to fears lest such a moving field should touch, and sink the ship.

At times the whole body of the pack, acted upon by north-east winds, would sweep gradually southward and towards the shoals and cliffs of Princess-Royal Island: indeed, at one time, the Investigator drifted twenty-four miles south in three days. They had fortunately laid hold of a large piece of ice which grounded upon the shoals westward of Princess-Royal Island; and under its lee the ship held on for security, whilst the rest of the ice swept by. Some idea of the occasional strain upon the ship, as well as the desperate position she was in, may be gleaned from the fact, that at one time she was in five fathoms water, and trusting for safety by holding on to heavy grounded ice by every available hawser in the vessel, amounting in the aggregate to a thirty-one inch hemp cable and a stream chain in addition, yet this was every minute expected to part, as the ice-pressure took her broad bow, or surged against her trembling sides.

Anticipating the worst that could occur, Captain M'Clure on these occasions placed a large quantity of provisions and fuel upon deck, told off the officers and men to their boats, and every man had his appointed place and duty in the event of a catastrophe; tents and warm clothing were also prepared, and every precaution was thus taken to save life, even if it were beyond human power to save the ship.

On the 27th September, the temperature being then at zero, and the ice, as they fancied, stationary, and the Investigator fixed in it, ten miles south of the Princess-Royal Island, preparations were commenced for housing the vessel over, and otherwise securing the crew from the intense cold and inclemency of a winter which was now wellnigh upon them. The officers were just congratulating themselves upon their escape from past dangers, and expressing thankfulness at having only lost thirty miles of latitude by the drifting of the pack, when a change of wind set it all again in motion. The 28th was spent in breathless anxiety, as, helpless in their icy trammels, they swept northward again before a south gale toward the cliffs of Princess-Royal Island. Its cliffs rose perpendicularly from the sea at the part against which the ship appeared to be setting, and as the crew eyed them for a hope of safety, if the good craft should be crushed against the rocky precipices, they could see no ledge upon which even a goat could have established a footing; and an elevation of 400 feet precluded a chance of scaling them. To launch the boats over the moving pack, as the ship sank, was their sole chance, and that a poor one, rolling and upheaving as it was under the influence of wind, tide, and pressure.

It is in such an emergency that discipline, and a certainty that each would perform unflinchingly his duty, as well as the innate good qualities of our noble seamen, are shown to the best advantage. Dastards would in such circumstances have deserted their ship; but the Investigators were made of different stuff. They knew too that One who is "strong to save" was watching over them, and they eyed the bleak cliffs, which in a few minutes might be frowning over their graves, with the calm courage of resolute men. A finer picture than such a scene presented can hardly be imagined, and it was one repeatedly exhibited during this wonderful voyage; but it would be an almost hopeless attempt to convey to the reader by mere description an adequate idea either of scenery so replete with the grim terrors of the polar regions or of the moral grandeur of self-devotion in the officers and men at such a crisis.

"It looks a bad job this time!" inquiringly remarked one of the sailors, as he assisted another old sea-dog in coiling down neatly a frozen hawser. "Yes!" was the rejoinder, as the other shaded his eyes from the driving snow, and cast a glance at the dark cliff looming through the storm; "the old craft will double up like an old basket when she gets alongside of them rocks!" The Investigator's hour was not yet come, however; and when within 500 yards of the rocks, the ice *coach-wheeled* her along them, and finally swept her past the eastern side of the islands.

After this water was never seen from the mast-head; yet onwards the ship drifted slowly, and on September 30th became again stationary in lat. 72° 50′ N., and long. 117° 55′ W., very nearly as far north as they had reached a fortnight before, and been drifted back from in the grip of the packed ice.

CHAPTER XI.

SEVERE PRESSURE AND DANGEROUS NIPS — FAREWELL TO THE SUN — HOUSING THE VESSEL — GOOD HEALTH AND SPIRITS OF THE MEN — FIVE HUNDRED POUNDS OF MEAT FOUND TO BE PUTRID — WINTER RAMBLES ON THE ICE — PERILS ARISING IN THESE TRIPS — AN EXCURSION TO VIEW THE NORTH-WEST PASSAGE — HARD LABOUR AND INSUFFICIENT FOOD — SUFFERING FROM THIRST — THE PASSAGE SEEN — CAPTAIN M'CLURE LOST FOR A NIGHT — RETURN OF THE PARTY TO THE SHIP — SUCCESS OF MEASURES TAKEN FOR THE HEALTH OF THE CREW.

DURING the first week in October a change of the moon occasioned spring-tides, which, of course, led to considerable motion in the ice; but that motion only manifested itself in the shape of severe pressure and *nips*, there not being sufficient space in the strait to allow the pack to drift either north or south. On the occasion of one of these nips, the Investigator was thrown much over to the starboard side, and lifted two feet out of the water by the ice pressing under her keel; every timber in the vessel cracked and groaned, and the bells began to ring as she surged and trembled under the shock. There needed no boatswain's pipe to bring all hands upon deck; and, in an October night, with the temperature 36 degrees below freezing-point, each man stood at his station, momentarily expecting a final catastrophe to the ship, and that they would be left upon the surface of the frozen strait, to fare as best they might.

Magnificent auroras lit up the heavens more than once about this time, and generally appeared most brilliant in the southern quarter. The pale sun swept, it is true, across the sky, in a daily-diminishing arch; but its rays had ceased to give warmth, and the tiniest crystals of snow withstood its power. Light, however, it still gave for a while; and all looked kindly upon a friend for whose speedy return they should soon have to pray.

The woollen housing was now spread over the vessel, and the curtains nailed down to the gunwale upon the northern side, to shield the men from the cutting blasts of that quarter; but to the southward every precaution was taken to enjoy the sun's light as long as possible. The fact of life and light being almost synonymous terms was deeply impressed upon the mind of Captain M'Clure; and to his constant remembrance of it we must in a great measure impute the extraordinary exemption of his crew from scurvy. They, as well as the officers, appeared now to be in the best health and spirits; and there were only two men upon the doctor's sick-list on the 6th October.

Every evening, after work was over, the after-part of the lower deck was converted into a temporary stage, on which the "clever dogs" of the crew performed, danced, sang, or recited for the amusement of those who were less accomplished; and roars of laughter and light-hearted jokes bore good evidence that neither nips, frostbites, nor hairbreadth escapes, preyed upon the spirits of any of the audience.

On examining some canisters of preserved meat, Captain M'Clure found, much to his chagrin, that no less than 500 lb. were so putrid as to necessitate throwing them overboard. This loss was mainly occasioned by fractures made in the tins when stowing the ship's hold in England. This diminution of resources was very deplorable; and it will be remembered that a boat-load of meat had already been lost when the ship was aground off Point Manning. The captain consoled himself, however, with the hope that a certain surplus, which the contractor had promised to put on board to cover such contingencies, would replace this unfortunate deficit.

Amongst the preparations made for the worst that could befall the ship, there is one which will strike every one as evincing carefulness and skill, and, at the same time, it will show how critical the position of the vessel must have been. This was the blasting with gunpowder, and the employment of manual labour, to remove all the hummocks and inequalities upon the surface of the ice upon one side of the vessel, in order that a smooth surface might be ready to receive her, as there was a probability of her being positively thrown upon the ice. Nothing can better bring home to our minds their position, or the cool way in which it was met, than the above fact; and to realise it, Captain M'Clure need hardly add, in his journal, that he despairs of being able to convey to us even a remote idea of the harassing anxiety he underwent whilst his vessel was settling herself in her icy cradle. "The crashing, creaking, and straining are beyond description," he adds; "and the officer of the watch, when speaking to me, is obliged to put his mouth close to my ear, on account of the deafening noise."

From the 10th of October the ice in and about the ship became fixed, although the whole body of the pack was still in some places detached from the shore, and moved slightly north and south with the tide. The work of housing over being completed, parties of men and officers began to stroll out for the purpose of acquiring some information of the neighbourhood, and going through the form of taking possession of their new discoveries; a ceremony which, though of no great importance when the acquisition was so entirely valueless, served at least to break the dreary monotony of such an existence as theirs had become.

The incidents which arose upon such rambles afforded something to talk about, too; and on some occasions unforeseen dangers added to the excitement of the journeys. One instance will serve to show how unpleasantly these parties of pleasure sometimes ended.

On a calm fine morning, with the temperature just 40 degrees below freezing-point, Captain M'Clure, Lieutenant Cresswell, Dr Armstrong, and Mr Mierching, with some seamen, started to visit the eastern side of the strait, and take possession of the land. The road at first lay over the broken and rugged pack; but they afterwards reached a belt of smooth ice of the present season's formation, and it carried them to a piled-up barrier of broken floe, formed where the new ice impinged against the heavy grounded hummocks which lined the coast.

The tide happened at the moment to have brought the two edges together with much violence; and the *lighter ice* (some feet in thickness, however) was turning up and rolling over, layer upon layer. Follow my leader was the idea of all the party; and away they rushed over the pile formed by the battling floes, cheering as they reached the land, and regardless of the fact that at turn of tide those very floes might part and cut off their retreat.

Every one put his best leg foremost to reach some high land seen in the interior, from the summit of which there would be a possibility of seeing into Barrow Strait, and thus connecting their discoveries with that of Sir Edward Parry in 1819. The seamen were left behind to construct a cairn upon a spot duly christened after the Illustrious Consort of Her Most Gracious Majesty; and the officers, after another two hours' hard struggle through deep snow, and over a difficult country, reached what was long afterwards remembered as Mount Adventure.

Although some 1400 feet above the level of the sea, the trending of the coast they were upon prevented their toilsome journey being rewarded by a view of the termination of Prince of Wales Strait upon the eastern side; but on the west there rose in the distance a headland, which appeared like the termination of Banks or Baring Land, with a blank space between it and the east side of the strait, which confirmed Captain M'Clure in his belief of a channel through, and made his companions exclaim that they saw into Barrow Strait. This point, however, the captain wisely decided upon placing shortly beyond all doubt or cavil, by exploring it hereafter with a sledge-party. Although traces were seen on the snow of bears, deer, foxes, and lemmings, they did not fall in with a single living creature; and the view they obtained of Albert Land was not such as to afford much promise of game—for vegetation, the great test of the presence of animal life in the far north, was exceedingly scanty, and little gladdened the eyes of our travellers beyond small patches of dwarf willow and moss.

"We had returned to the shore," says Captain M'Clure, "and were following our track back to the ship, anticipating the pleasure of a good dinner after a twenty miles' walk, when, upon coming to where the junction of the land (or fixed ice) and sea floes took place, we beheld a separation of fifty yards of clear black water! Our feelings are easier to be imagined than described; nearly five miles from the vessel, a polar night closing in, and the only provision amongst the whole party was a solitary tin of preserved meat which had been issued to the men for their dinner, but had now become so solidly frozen as to defy both their knives and teeth."

Just before dark, a point a few miles to the southward was observed, which gave some promise of being connected with the sea-floe by a block or barrier of ice. Towards it the fatigued party struggled, over very rugged and slippery ice. Every now and then one of their party would experience a severe fall into some deep cleft, or over some huge hummock; and then, thoroughly jaded, they would sit down and feel inclined to drop off into a sleep from which they would never have awakened in this world. Captain M'Clure, however, was aware of this danger, and his voice aroused them to exertion. After firing muskets for some time to attract attention, they were rejoiced to see rockets and guns discharged from the ship. It told them that those on board were taking measures for their rescue; and, meantime, they continued to indicate their position to the ship by firing at intervals so long as their ammunition lasted. About half-past eight a light was seen evidently approaching upon the sea-ice. A shout of delight, responded to as heartily, rang through the black stillness of a polar night. Then came the anxious hope that the people from the ship had brought a boat with them—for without it aid was out of the question.

Even in such a moment the sailors' light-heartedness did not desert them; for when one of the party exclaimed that "the ship had fired another rocket!" "Ah!" another observed, "I wish they would fire a Halkett's boat * at us!"—a wish in which assuredly all cordially joined.

The relieving party at length approached within hail, upon the opposite side of the lane of water; and, worn out as Captain M'Clure's party was, all listened with indescribable anxiety for the answer when, to the momentous question put by the leader, "Have you a boat with you?" there was a pause, in which the writer has been assured one could have heard a pulse beat, and

* Halkett's boats are the ingenious invention of Lieutenant Peter Halkett, of the Royal Navy. They are made of India-rubber, and, being inflated with air, are very portable, and highly useful upon arctic service.

then came across the darkness—"No! we did not know you wanted one." Captain M'Clure sent them back immediately to the ship for the Halkett's boat, and, meantime, aided by his officers, he exerted himself to keep the men from falling asleep. Happily the party, on its way to the Investigator, was met by another conveying boats, and the two returned with all speed to the water, which had now become covered with bay-ice nearly an inch thick.

Mr Court, the master of the Investigator, was just the man to meet such difficulties as now lay in the way of relieving his shipmates; and in spite of bay-ice, and current, and moving ice, by midnight all the party were safely ferried across, and on their way to their ship. "I cannot," says Captain M'Clure in his journal, "speak too highly of these excellent little boats, or of the ingenuity of the inventor, as without them my large party would have had to endure the rigours of an arctic night, without clothing, tents, or provisions, and the consequences of this might have been very serious."

By four in the morning the travellers had partaken of a substantial meal, and retired to their beds heartily tired after eighteen hours' exertion, and grateful for so fortunate a termination to their adventure.

From the 10th to the 21st of October, preparations were made to despatch a sledge-party to the northward to reach Barrow Strait, and to positively assure themselves of the fact of their having discovered a North-West Passage. Even had they been ready to start at once, it would have been necessary to give time for the ice to form sufficiently to insure the ship from being blown away with the drifting pack whilst the party was absent—an accident which experience has shown to arctic navigators might occur up to a late date in October. Just at this time a remarkable rise of temperature to 24° *plus* of Fahrenheit, from 2° *minus*, whilst the wind was blowing fresh from *north-east*, would seem to indicate that the winter of the region in which the Investigator was frozen in, is modified by the warm air from the open water of Barrow Strait, as well as from that forced up by southerly winds from the American continent. This sudden change, however, was far from pleasant to the crew, for they had all put on their winter clothing, and had begun to shut up the ship, ready to resist the rigour of winter; a momentary rise of this nature, therefore, only created discomfort, and was of too transitory a nature to be beneficial. Indeed, the men voted warm weather in the middle of October a nuisance; and the old hands, with a knowing shake of the head and copious expectorations of "'baccy juice," warned the novices against "being fools enough to pull their clothes off on account of a bit of sunshine, for perhaps in an hour's time Zero would be about again." *Zero*, it must be observed, was invariably referred to as a veritable foe having an actual existence, and was to be combated as they would do the Arch-Enemy.

A landing was now effected on the islands named after Her Royal Highness the Princess-Royal; but they offered nothing remarkable beyond some ancient Esquimaux graves and fox-traps. Traces of animals were, as usual, numerous, and excited as much interest in the minds of our navigators as human footsteps did in that of Robinson Crusoe. Those who could appreciate the possible contingencies of arctic exploration were keenly alive to the importance of procuring game of some sort, to eke out the resources of the ship, and keep the crew free from the ravages of scurvy during the forthcoming winter. Even as a question of mere rations, apart from the desirability of obtaining fresh food for the crew, the subject forced itself upon every one's attention, for the preserved meats were constantly found to be in a putrid condition; and between the 12th and 18th of October no less than *four hundred and twenty-four pounds* were thrown overboard as unfit for food—much to the regret of Captain M'Clure, who was thus early warned how carefully he should be obliged to husband his resources, to carry his crew through their enterprise successfully.

As yet, however, no reduction in the allowance took place; for the leader of that gallant ship's company knew that, when the time came to render a straitened allowance *actually necessary*, his officers and men would cheerfully and manfully submit to the privation.

October the 21st, 1850, came in with a temperature ranging a little below zero, light winds, and an overcast sky. The ice of the strait appeared to have remained stationary during the last spring-tides, and the usual polar accompaniment of strong gales; Captain M'Clure thereupon determined to start for Barrow Strait with a sledge manned with six men, and commanded by Mr Court, his active and indefatigable master, aided for a while by a fatigue party of men under Mr Wynniatt (mate) and Dr Armstrong, and to leave the ship to the charge of Lieutenant Haswell. Nothing can be more delightful than the terms of warm praise in which Captain M'Clure speaks of all the officers, when upon the eve of parting from them for a service not unattended with some peril. Hearty were the cheers, and Well-fare-ye's! on either side, as the little sledge-party bade good-bye to ship and companions, and plodded on their lonely way, to bring back one day to their shipmates the most interesting intelligence ever told to the hundreds who have devoted health, strength, and energy to the problem of a North-West Passage.

The headlong zeal of the excited crew upon the sledge soon received a lesson in patience from the rugged and broken pack, by the repeated capsizing of the sledge, and its eventual fracture beyond all temporary repair. There was nothing left for them but to send Mr Court back to the ship for another sledge, whilst the party pitched their tent, and slept under canvass upon the frozen ocean.

The Investigator had left England but little prepared for extensive sledge-work, and with few, if any, improvements upon the system of sledge-travelling originally laid down by Captain Sir James Clark Ross. In all her sledge-parties there was consequently a greater amount of hardship and privation than in those of the expeditions under Captains Austin, Kellett, or Belcher, who each improved upon their predecessors' experience. We find, for instance, that at the close of the first day's journey, the truly frugal meal of Captain M'Clure and his men was a pint of tepid water apiece, into which a little oatmeal was thrown; after which they retired to their sleeping bags, to rest as best they might with a temperature of 6° *minus*. On 22d October, the new sledge having joined them, the party proceeded to the northward, working over alternate patches of rough and smooth ice, until the night came on, and it became too dark to see their way. The tent was then pitched, and supper prepared; but such a supper! one pint of melted snow and a piece of frozen pemmican! Hunger, however, sweetened even this meal; and, tired and cold, they got into their frozen blankets and fell asleep, whilst an October snow-storm rolled over their frail canvass tenement. Next morning before daybreak the cook of the day was roused, and his culinary powers were called into play under the trying circumstances of a temperature of 32° below freezing-point. Some water was warmed sufficiently for a preparation of chocolate to be dissolved in it; and a pint of this tepid beverage being given to each man, together with a biscuit and a half apiece, the party again manned the drag-ropes of the sledge and proceeded to the northward.

After some difficulty in crossing ridges of broken ice—the *hedgerows* of an arctic landscape—they reached vast fields of smooth ice of the present season's formation; and here an obstacle of a fresh nature awaited them. The autumnal snow had accumulated heavily upon the surface of these young ice-fields, and, weighing them down, caused the sea-water to flow through sufficiently to render the under part of the snow almost as tenacious as clay. The fatigue of hauling two hundred pounds apiece through such a route was excessive; but the gallant crew strained every nerve, and the distended veins and large drops of perspiration (freezing on the faces of the men) told how well they were working. Unfortunately no water could be had to appease their thirst—they might as well have been labouring on the great Sahara; for every handful of snow which they thrust into their parched mouths augmented rather than assuaged their sufferings, as it contained more or less of the salts of the sea-water which rendered the surface of the floe wet and tenacious. About noon, one of the best men of the party became perfectly exhausted, and two others were frostbitten. Captain M'Clure then stopped to give them the noonday meal of cold water and frozen pemmican—but the latter they did not taste, for thirst had quite overcome hunger; and when they had drunk all the water that the allowance of fuel for the day would thaw, they again trudged on till dark, when, as on the previous evening, the tent was pitched, and their rough meal and rougher bed prepared.* Pipes were then lit; and whilst some of the men repaired their torn mocassins and seal-skin boots by candle-light, the Captain read them a tale out of 'Chambers's Miscellany,' until at last his tired companions fell off one by one, to forget their sufferings, into the land of dreams.

"October 24," says Captain M'Clure, "was not so cutting a day, the thermometer having risen to 5° Fahr.; I walked ahead whilst the sledge was packing, ascended a point of land a hundred feet above the level of the sea, and observed distinctly that the eastern shore of Prince of Wales Strait trended now far away to the

* This system of sledge-travelling has in more recent expeditions been much improved upon, and our men have made very much longer journeys with comparative comfort, and abundance of food.

eastward, whilst that of the western coast (which we were upon) preserved its northerly direction. The point whereon I stood appeared to be the most contiguous to the opposite shore, and the breadth across about fifteen miles; beyond me, the shores of the strait evidently began to separate. This encouraged me in the hope that we were on the point of reaching Barrow Strait; and seeing a hill at what appeared a distance of 12 miles due north of my position, I returned to the sledge, and pointed it out to the crew as a cape from whence we should see that long-wished-for sea."

Every man now dragged with a will, in the hope of reaching that night the end of his journey; but after seven hours' toilsome labour, the tantalising cape still retained its original position, and they seemed not a mile nearer to it. Captain M'Clure then saw that he had been much deceived in its apparent distance owing to the clearness of the atmosphere, and that thirty miles was a nearer estimate than twelve, of the probable length of their march. After a night's rest and another hard day's work, they were still two miles off the cape, when night closed in and obliged them to halt and encamp. Though disappointed in not yet having sighted Barrow Strait, they were all much cheered by the multiplying proofs around them of its close proximity. Away to the north-east they already saw that wonderful oceanic ice which Sir Edward Parry so well described in his memorable voyage to Melville Island in 1819—ice which they had left behind them directly they passed the southern entrance of Prince of Wales Strait, and which they now again found at its northern extremity. Great hills and dales of blue crystalline sea-ice rolled on before them in the direction of Melville Island; and it required more than ordinary sanguineness of disposition to suppose they ever should navigate the old Investigator through such a sea; yet, to have heard the party talk, the feat appeared certain of accomplishment—all things seemed possible to men who had already mastered so much. By an observed meridian altitude of the star Capella, the latitude on October the 25th was 73° 25′ N., and this was the first and only observation they had been able to obtain since quitting the ship.

The morning of the 26th of October 1850 was fine and cloudless; it was with no ordinary feelings of joy and gratitude that Captain M'Clure and his party started before sunrise to obtain from the adjacent hill a view of that sea which connected their discoveries with those of Sir Edward Parry. Ascending a hill 600 feet above the sea-level, they patiently awaited the increase of light to reveal the long-sought-for North-West Passage from the Atlantic to the Pacific Oceans.

As the sun rose, the panorama slowly unveiled itself. First, the land called after H.R.H. Prince Albert showed out on an easterly bearing; and from a point since named after the late Sir Robert Peel, it evidently turned away to the east, and formed the northern entrance of the channel upon that side.

The coast of Banks Land, on which the party stood, terminated at a low point about twelve miles further on —thus forming a part of, and connecting itself with, that land, the loom of which had been so correctly reported and so well placed on our charts by Sir Edward Parry's expedition thirty-odd years before. Away to the north, and across the entrance of Prince of Wales Strait, lay the frozen waters of Barrow, or, as it is now called, Melville Strait; and raised as our explorers were, at an altitude of 600 feet above its level, the eyesight embraced a distance which precluded the possibility of any land lying in that direction between them and Melville Island.

A North-West Passage was discovered!* All doubt as to the existence of a water communication between the two great oceans was removed; and it now alone remained for Captain M'Clure, his officers and men, to perfect the work by traversing the few thousand miles of known ground between them and their homes.

The feelings of Captain M'Clure and his companions may be easily understood when we remember what they had gone through to earn this success, and how the hand of the All-powerful had borne them through no ordinary dangers in their gallant efforts; but no arrogant self-estimation formed part of the crowd of tumultuous feelings which made their hearts beat so high, and never from the lips of man burst a more fervent *Thank God!* than now from those of that little company.

* The subsequent recovery, by Captain Sir Leopold M'Clintock, of the relics and records of the expedition under Sir John Franklin, prove that his ill-fated crews, coming from the Atlantic, did in the year 1848 perish on the coast of America, at or about the mouth of the Great Fish River. That position had been long known to communicate directly with the Pacific Ocean by way of Behring Strait. The priority of the discovery of the North-West Passage clearly, therefore, belongs to Franklin's expedition; but the credit of discovering two other water communications, ice-choked though they be, on either side of Banks Land, between the waters of the Atlantic and Pacific, belongs to Sir Robert M'Clure.

And we feel that they had reason to be proud as well as grateful, when we call to mind the time, the money, and the men which England had previously lavished, without success, on the discovery of this great geographical problem.

Franklin and his heroic followers had, indeed, not then been found; but, in seeking them, a great problem had been unravelled, and Captain M'Clure felt that, even should he be so unfortunate as never to discover the missing expedition, he nevertheless would not return to his country with empty hands.

The position of Mount Observation, from which the important discovery had been made, was ascertained to be in latitude (observed) 73° 30′ 39″ N., longitude 114° 39′ W., and by lunar 114° 14′ W. Pushing on to the extreme northern entrance of the strait, the travellers encamped that night on Cape Lord John Russell, and cheered lustily as they reached the shores of Barrow Strait. A mimic bonfire, of a broken sledge and dwarf willow, was lighted by the seamen in celebration of the event; and an extra glass of grog, given them by their leader, added to their happiness.

The question of a north-west passage in this direction being thus placed beyond all doubt, the rapid fall of temperature warned Captain M'Clure that he should return to his ship without delay, and terminate the trials the whole party were exposed to every night. Their fur robes were frozen into a solid mass, which could only be thawed by the men lying upon them for some hours; the blanket bags were so stiff from the same cause as to stand erect; the clothes, caps, whiskers, and beards of the party were constantly frozen together, and required to be thawed inside the tent before they retired to rest; and when their clothes were taken off, they had to be placed under their bodies that they might not freeze again. In fact, the hardships and discomforts to be endured in consequence of the lateness of the season, although common to arctic travellers, would, if minutely described, appear almost fabulous to others.

From Point Lord John Russell the coast of Banks Land was seen to trend away to the westward, and increase in boldness of outline and altitude. Much vegetation for this latitude was observed, and numerous traces of animals, such as the deer, hare, and ptarmigan, as well as of their destroyers the fox and the wolf; but not one of the animals themselves was seen. A large cairn was constructed, a due record of the visit of the party placed therein, and then, in the teeth of a S.E. gale, they commenced their return to the Investigator.

The return journey might have ended seriously for the leader of the party. On the 30th October, at 2 P.M., having seen the Princess-Royal Isles, and knowing the position of the Investigator from them, Captain M'Clure left his sledge, with the intention of pushing for the ship and having a warm meal ready for his men on their arrival. When still six miles from the ship the night overtook him, and with it came a dense mist accompanied with snow-drift, which rolled down the strait, and obscured every object. Unable to see his road, but endeavouring to preserve a course by the wind, M'Clure continued to hasten on, until repeated and heavy falls amongst the broken ice warned him to desist or incur the additional peril of broken limbs. "I now," he says, "climbed on a mass of squeezed-up ice, in the hope of seeing my party, should they pass near, or of attracting the attention of some one on board the vessel by firing my fowling-piece. Unfortunately, I had no other ammunition than what it was loaded with; for I had fancied when I left the sledge, that the two charges in the gun would be all I should be likely to require. After waiting for an hour patiently, I was rejoiced to see through the mist the glare of a blue light, evidently burnt in the direction in which I had left the sledge. I immediately fired to denote my position, but my fire was evidently unobserved, and, both barrels being discharged, I was unable to repeat the signal. My only hope now rested upon the ship answering; but nothing was to be seen, and although I once more saw, at a greater distance, the glare of another blue light from the sledge, there seemed no probability of my having any other shelter for the night than that the floe afforded. Two hours elapsed. I endeavoured to see the face of my pocket-compass by the light of a solitary lucifer-match, which happened to be in my pocket; but in this hope I was cruelly disappointed, for it fizzed and went out, leaving me in total darkness. It was now half-past eight; there were eleven hours of night before me, a temperature 15° below zero, bears prowling about, and I with an unloaded gun in my hands. The sledge-party might, however, reach the ship, and, finding I had not arrived, search would be made and help be sent; so I walked to and fro upon my hummock until I suppose it must have been eleven o'clock, when that hope fled likewise. Descending from the top of the slab of ice upon which I had clambered, I found under its lee a famous bed of soft dry snow, and, thoroughly tired out, I threw myself upon it and slept for perhaps three hours, when upon opening my eyes I fancied I saw the flash of a rocket. Jumping upon my feet I found that the mist had cleared off, and that the stars and aurora borealis were shining in all the splendour of an arctic night. Although unable to see the islands or the ship, I wandered about the ice in different directions until daylight, when, to my great mortification, I found I had passed the ship fully the distance of four miles." Retracing his steps, Captain M'Clure reached the Investigator on the 31st October very tired, but otherwise none the worse for his rough and dangerous exposure to a winter's night in 73° north latitude. A few hours afterwards the sledge-party arrived under Mr Court; and great was the joy on board the ship, and hearty the congratulations upon their safe return, and the glorious news they brought.

Nothing, I fancy, can better bring home to the comprehension of the uninitiated in arctic sledge-travelling, the severity of the labour undergone by officers and men employed upon duty of that nature, than the following extract from Captain M'Clure's private journal—and similar ones might be found in those of many other officers:—"The weight brought back to the ship upon the sledge (after an absence of nine days) was 793 pounds, being an *increase* upon what we started with of upwards of 100 pounds. This was occasioned by the accumulation of ice upon the furs, tent, blankets, and sledge, in consequence of the vapour thrown off by our bodies and cooking apparatus condensing and freezing upon every article which it came in contact with. And, strange as it may seem, the whole consumption of food during nine days amongst eight men, independently of chocolate and spirits, amounted but to eighteen pounds of pemmican, thirty-one pounds of biscuit, and two pounds of oatmeal —a consumption almost incredible, and only to be accounted for by the crew being every night too exhausted, after their day's exertion, to care for anything else but water; and this article was not to be obtained without thawing it, and the allowance of fuel would only admit of each man receiving daily five gills to drink—namely, half a pint at breakfast, a gill at dinner, and half a pint in the evening."

On this, however, they had worked cheerfully, and accomplished an average of twenty miles per diem—a feat which it is but right to say only the discovery of a north-west passage could have carried the men through; for although Lieutenant (now Commander) Mecham has in later years far excelled Captain M'Clure's journey with respect to distance accomplished, it was only by carefully feeding up and nursing the strength of his men that he at the same moment enjoys the honour of having won the palm in daily distance accomplished from such men as Captains Richards and M'Clintock.

During the absence of their captain, the officers of the Investigator had been far from idle. Upon the adjacent shores of Prince of Wales Strait they succeeded in shooting a fine herd of musk-oxen, consisting of three bulls, a cow, and a calf, yielding a supply of 1296 pounds of solid meat. The moral effect of the discovery that such a quantity of fresh food could be found near a place where they were frozen up, was very beneficial upon the minds of the crew, and added materially to the feeling of general confidence with which they prepared to meet the coming winter.

The ventilating tubes to the lower deck were now fitted, to force out by a current of pure but cold air the heated and deleterious vapours generated between decks by a number of men living in so confined a space. The last winter housings were finally secured down, and a winter school-room established, to which thirty pupils immediately repaired to learn to read and write; and by the 11th of November the Investigator was ready to bid the bright sun good-bye.

The day was cloudless, the temperature down to 26° *minus*, and one uniform sheet of snow and ice spread on every side, over land and sea. Winter had set her seal upon that silent strait, and but for the rocky buttresses of Princess-Royal Island frowning over the floe, or the dark cleft of a ravine upon Banks Land, it was not easy to detect the line of demarcation between earth and water. Towards noon the bright edge of the upper limb of the sun rolled slowly along the southern horizon, and bade them adieu for eleven long weeks; the dreary night of a polar winter had commenced. Between the 2d of November and the 2d of December the new floes were found to have increased in thickness ten inches and a half, the last measurement making them 2 ft. 6½ in.; yet little if any snow could be found on them for drinking purposes—a serious inconvenience, arising from the weight of the snow-drifts breaking down the weak ice, causing the sea-water to flow through, and the sludge so formed would, in so low a temperature, rapidly become a part of the solid floe. In this manner the ice that covers the arctic sea accumulates more rapidly in the early winter by the deposit upon its surface than by the congelation of the water beneath.

During the first fortnight in December the temperature of the external air ranged from – 23° to – 37° Fahr., whilst internally between decks from + 40° to + 50° was about the average.

It had been Captain M'Clure's great anxiety to insure warmth on the lower deck, without having an accumulation of wet arising from the natural condensation of the heated internal atmosphere against the cold surface of the sides and deck; and he succeeded (as had been done in Captain Austin's expedition) in securing this desirable end, by fitting ventilators and clearing the lower deck of men for the major part of the day, so that a free current of air should circulate for a while throughout the vessel. By these means he secured the health of his men to a degree previously unprecedented upon arctic service; and they duly appreciated the forethought and care thus bestowed upon them, and in the following season they resumed their duties as if fresh from England, and enabled Captain M'Clure to achieve a still more perilous voyage than the one already accomplished in the year 1850.

The minute details of the daily events of an arctic winter have been so often described that it is unnecessary to recount them. They consist of remarks upon decreasing warmth and daylight, varied sometimes by a notation, that on such a day there was an unusually brilliant aurora borealis, or a great frequency of shooting-stars. The arctic fox, of course, came as usual to visit the new intruders upon his domains, but only to be trapped and have his snow-white fur packed up to ornament the neck of some fair lady at home. But it was now for the first time noted, in so high a latitude, that the arctic raven, the hardiest of the feathered tribes, was seen in the depth of the dark season to flit through the cold and sunless atmosphere like an evil spirit, his sullen croak alone breaking the silence of that death-like scene. No one shot any of these birds of ill omen; and they seemed to feel they were secure.

Christmas came at length, with all its hallowed recollections; and it was kept on board the Investigator, as it is on board of a man-of-war in every part of the world, in cheerfulness and in good-fellowship. The Captain's table groaned under good cheer. There was beef from the Sandwich Islands, which had been kept in a frozen state for six months; there was veal (of musk-ox) from the shores of Prince of Wales Strait; there was mincemeat from Old England, splendid preserves from the Green Isle, and many a dainty dish from Scotland. Each one talked of home; they calculated the hour when, allowing for the difference of time, those most dear to the talkers would be going to church, to dinner, to ball, or to bed; and an honest manly hope was expressed that, one of those days, they might yet be there to see and share in happiness, in their estimate of which distance naturally "lent enchantment to the view."

Thus closed the year 1850. The Investigator that

day had only one man ill, and he was one who had concealed the fact of his being in delicate health when joining the ship at Woolwich. "Every credit," says Captain M'Clure, "is due to the medical officers, Drs Armstrong and Henry Piers," for their unremitting attention to the health of the men; "and nothing could be more satisfactory than the state of the vessel, her crew, and her resources on this day,"—the last of the year 1850.

CHAPTER XII.

NEW YEAR'S DAY, 1851—RELATIVE POSITIONS OF THE DIFFERENT EXPEDITIONS—THE INCREASING COLD RELIEVED BY THE DAILY AUGMENTATION OF SOLAR LIGHT—DEER AND PTARMIGAN SEEN IN THE DEPTH OF WINTER—THE THEORY OF ANIMAL MIGRATION IN ARCTIC REGIONS SUBVERTED—THE RAVEN LEAVES THE SHIP—RETURN OF THE SUN—RAMBLES ON THE ICE—REVIVAL OF HEALTH AND SPIRITS—WINTER SPORTING—PREPARATIONS FOR SLEDGE-PARTIES IN SEARCH OF FRANKLIN'S EXPEDITION—DEPOTS ESTABLISHED TO SECURE THE SAFETY OF THE TRAVELLERS—DEPARTURE OF THE VARIOUS PARTIES—HARDSHIPS ENDURED BY SLEDGE-CREWS IN HIGH LATITUDES IN SPRING JOURNEYS—THE ZEAL AND COURAGE OF THE SEAMEN—THE SCENE OF THEIR LABOURS COMPARED WITH SOUTHERN LATITUDES—THE POSITION OF THE INVESTIGATOR IN 1850—MURDER OF LIEUTENANT BARNARD BY NORTH-WEST INDIANS.

THE winter of the year 1850-51 was a remarkable one in the arctic regions. On the side of Baffin Bay, a naval expedition, consisting of Her Majesty's ships Resolute, Assistance, Pioneer, and Intrepid, manned by 180 officers and men, had pushed into the ice of Barrow Strait, until caught by the winter under Griffith Island.* Not many

* These vessels were commanded by Captain T. H. Austin, C.B., Captain Erasmus Ommanney, Lieutenants-Commanding John B. Cator, and Sherard Osborn.

miles from them, in a small bay in North Devon, two handy little brigs under Captain W. Penny, a noble specimen of the merchant sailor, lay securely housed in, manned by fifty sailors chosen from the hardy whaling-crews of Aberdeen and Peterhead. Close to these last vessels, an English yacht was wintering, under the command of that veteran, Admiral Sir John Ross. Three-score and ten years had not quenched in him that strong love for hardship and adventure which seems the only assignable motive that can induce men to continue to follow the hazardous career of an arctic navigator. But on the occasion of which I am writing, a nobler and higher purpose carried that aged officer to the frozen regions.

Thus, on the *eastern* side of the unknown waste which lay between Banks Land and Griffith Island, we have these seven vessels securely wintering, and prepared, with no small zeal, to push out their sledges directly the daylight and temperature would admit of it; while on the *western* side the Investigator alone, far from all communication with either savage or civilised man, was flying her pendant with as much pride and confidence as if the solitude into which she had boldly pushed was the spot, of all others, her gallant crew would most desire to pass their New Year's day in.

All the various expeditions had good cause for contentment, and reason to be grateful to God; for their ships were secure, the ice was stationary, and though it was dark, and cold, and cheerless without board, within there was warmth, food, good-fellowship, and perfect health. Far different was the position of another expedition which had left its home on the same mission of carrying rescue to Franklin.

In 1850 an American party, under Lieutenant Commander de Haven, had, in two schooners, pushed up to Griffith Island at the same time with the English ships; but, being unprepared for wintering, the Americans tried to escape the grip of an arctic winter. Under sail, they bore up for Baffin Bay; but the rapidly-forming ice seized upon their ships, and, cradled in it, the Reserve and Advance drifted whither it listed. Death threatened them in every shape, their vessels groaning under pressure at one time, and then tossed about by broken ice in the fury of midwinter gales. Scurvy broke out amongst the crews. The vessels were not liberated from their icy fetters until the pack had swept them, through a hundred dangers, into Davis Strait. Well might our English seamen congratulate themselves on the immunity they enjoyed from the severe winter sufferings of their American coadjutors, and the still more hapless position of the crews of the Erebus and Terror; for some of those poor fellows might possibly have been still alive on that New Year's day. Alas! who shall tell how those sad hours of their last new year were passed by those gallant men!

It is necessary to the clear comprehension of the voyage of the Investigator, that the reader should bear in mind the relative positions of the other ships as I have given them; and that, at the same time, Dr Rae was wintering on the shores of the Great Bear Lake in North America, ready to start, with boat and sledge, northward immediately that the weather would allow him.

The Investigator's New Year's day was a happy one; many a delicacy long and carefully hoarded was produced on the table, at which all the officers and their captain met; and not the least remarkable of these dainties was a quarter of mutton which had been procured at the Sandwich Islands in the previous July—a pretty good proof of the preservative qualities of frost. On this day there was still but one man on the sick list, and the crew now felt that the most trying portion of the winter would soon be past, for with every returning day the sun was again approaching the horizon, and, slowly though it was, still the twilight *was* augmenting daily. Light was what they, as well as all others who have wintered in the north, most sighed for. The cold, however intense, is robbed of half its terrors if there is light to enable the arctic navigator to see around, and allow him to take his walk, or, gun in hand, seek for game.

The darkest period of an arctic winter is from about the 10th of December to about the 6th of January, whereas the lowest temperatures usually occur afterwards, in February, when there is considerable twilight—and, in the latitude of the Investigator's winter-quarters, some hours of sunlight. This merciful dispensation of Providence, by which the most rigorous temperature of the pole occurs after the period of total darkness has passed away, is one amongst the many which strikes the least observant visitor to those regions.

From the 9th January 1851 to the 16th, was the coldest period registered on board the Investigator; but there was tolerable light then from 9.30 A.M. to 2.30 P.M., so much, indeed, that at noon on the 16th the only star whose light was not quenched by the twilight was the bright star Arcturus. The spirits of the men rose, in spite of the thermometer showing 40° to 50° below zero of Fahrenheit! What cared they for quicksilver being solid, any more than for the solidity of the surface of the sea over which they strode? No, their health and spirits were good, they could see that the sun was coming back, and did it not promise them all they wanted?—summer thaws, open water, fresh adventures, the discovery of the Erebus and Terror, and then huzza for England!

Firmly believing the old-fashioned story of the annual migration of animal life in arctic America, it never entered any one's head that, during all this darkness and cold, there was abundance of fresh food close at hand. The discovery was an accidental one. Early one dark morning in January, a man named John Eames was walking upon the floe, some distance from the ship, and saw pass close by him a small herd of reindeer, trotting quietly towards Princess-Royal Islands. Had the ghost of his grandfather suddenly appeared to him upon the floe, John Eames could not have been more astonished; for he, like every one else, confidently believed in every living creature having gone to more favoured climes to the southward, until the summer should return. The news quickly spread; appetites sharpened; and sportsmen issued forth to slay venison. But the deer were not to be found, although they found some ptarmigan. These discoveries gave rise to much astonishment; how birds could exist in such a temperature, with the land covered deeply in snow? and with the soil, wherever it happened to be exposed, frozen so hard as to destroy iron tools in attempting to loosen it! There was, however, no doubt of the existence of both bird and beast in the neighbourhood, and doubts naturally suggested themselves of the correctness of the theory of animal migration laid down by that eminent naturalist and traveller Sir John Richardson, as well as by the late Admiral Sir Edward Parry; and Captain M'Clure, in his journal, says, "It is pretty evident that, during the whole winter, animals may be found in these straits, and that the want of sufficient light alone prevents our larder being stored with fresh food."

Subsequent observation throughout the arctic zone has completely overthrown the idea that the reindeer, musk-ox, or other animals inhabiting the archipelago of islands north of America, migrate southward to avoid an arctic winter. Throughout Banks Land, Melville Island, Bathurst and Cornwallis Land—in short, wherever British seamen have wintered of late years—there have been found indubitable proofs of the reindeer, bear, musk-ox, marmot, wolf, hare, and ptarmigan—in other words, all the fauna of those climes—wintering in the latitudes which they frequent during the summer.

January closed in with strong gales of wind from the westward; and, on one occasion, the wind veered to S.W. and blew with such violence from the more genial regions of America, that the temperature rose from —32° to —15° of Fahrenheit,—a change which, however pleasing, could not be thoroughly enjoyed for the snow-drift, which was too heavy for a soul to venture outside the friendly shelter of the woollen housings. An incident characteristic of life in the far north happened at this time. A raven that had haunted the ship during the past period of cold and darkness disappeared; and its departure was quite an event, something for the men to remark and talk upon, and its society was more missed than the loss of a more pleasing pet would have been elsewhere.

"The absent bird was a loss," says the gallant captain of the Investigator, "which we all felt; it had been the only creature that appeared as isolated as ourselves, and a mutual confidence had been established between us. The raven used to visit the ship unmolested except by the dog—who appeared to know the bird as well as we did, was always on the look-out for its visit, and went out to meet it occasionally. The dog would run at Ralpho; but he would hop over his head, and resume his occupation at the dirt-heap, keeping an eye, however, all the while upon the dog, and uttering a harsh croak occasionally, as if enjoying the fun of tantalising him."

On February the 3d the glorious sun rose again, after having been absent since the 11th of November. Eighty-four days of twilight and darkness! Few but the dweller in those high latitudes can understand the joy with which the return of that bright luminary was hailed; and the congratulations exchanged upon having been spared to rejoice again in the blessed sunlight, were mingled with heartfelt aspirations for the future.

Officers and men were every day extending their walks. Many a party was made up to Princess-Royal Island, each being sanguine of bringing back a well-filled game-bag; but the evening often saw the sportsmen returning unsuccessful and tired, with no other consolation than that of having seen at a distance some solitary wolf, and —upon the principle of "where there are bees there must be honey"—they strongly maintained that those creatures proved venison to be in the neighbourhood, and this venison might yet be theirs. Deer-stalking in a temperature of 60° below freezing-point, when all the country is buried in snow, and the sportsman stands out in strong relief upon the snowy landscape, is seldom remunerative to the larder—but it has the merit of giving occupation to minds pressed down by the canker of monotony; and the officers could smile and enjoy the marvellous tales brought back by the men of the number of miles they had walked, the quantity of game seen, or the size of reindeer footprints upon the snow, and at the excellent reasons for neither flesh nor fowl being found in their game-bags.

That it did not become warm directly the sun rose, was vexatious to those not gifted with patience; and many sighed at seeing the thermometer on February 21st only registering —44° in the shade, and that the sun-rays playing upon the bulb of the instrument only raised it to —28°, or 60 degrees below freezing-point! Outdoor sports now commenced; and to see the heavy falls the men experienced in their thick winter clothing and cloth snow-boots, whilst playing rounders upon the ice, an observer might have wondered how they escaped fractured bones and broken heads. Appetites that had failed now began to return, pale and yellow faces again to recover their ruddy and sunburnt colours; and long discussions already arose as to how Jack would spend his money when he arrived in England—an anxiety which in every clime weighs upon his mind when nothing else will.

Arrangements connected with the travelling operations of the coming spring were now entered upon; and although the present thickness of the ice in Prince of Wales Strait gave no promise of an early disruption, still Captain M'Clure determined, before the sledges left the ship, to establish such a depot, and place such means on the islands, as should render the sledge-parties independent of the ship, in the event of the ice breaking up and sweeping the Investigator north or south before their return. Early in March, therefore, a whale-boat was carried on sledges, with much labour and difficulty, to Princess-Royal Island, and a depot established of three months' victualling for the entire crew; so that, should the ship even be destroyed during the summer operations of 1851, a portion, if not all of the crew, might escape to the Mackenzie River or Barrow Strait, where some of Captain Austin's expedition would be met with. With this depot of provisions a record was placed, stating by whom and why it was established, and beseeching any parties from other ships that might visit it, to consider the provisions as sacred, and only to be touched upon the most urgent necessity.

These precautionary measures taken, the attention of leader, officers, and men was turned to the equipment of the sledges for their journeys over the ice in search of Franklin, as well as to the expediency of communicating the Investigator's position to any ships that might be in their neighbourhood.

Early in March the temperature in the sun rose to +10°, and heavy gales, with much snow, indicated the breaking up of the winter season; and as the action of the tides had already occasioned numerous cracks in the ice, Captain M'Clure landed another boat upon the eastern shore of the Strait, to enable the sledge-parties to retreat upon the depot, in case the ship should be carried away by the ice.

April brought rapid increase of sun, light, and heat. Embankments of snow were removed, daylight admitted below, and the walks of the officers still more extended. Game was sometimes seen, and ptarmigans occasionally shot; but there was too much work to be done connected with re-stowing and examining the state of the provisions in the vessel, and equipping the sledge-parties, to allow of any systematic plan of procuring fresh food being pursued during this their first winter.

On the 17th of the month the temperature, which had risen steadily, stood at +38° in the sun, and the floe around the ship became studded with pools of water—formed rather, however, by the tide forcing itself up the cracks and weak points in the packed ice, than by any action of the sun upon its surface. An early summer was naturally anticipated; and profiting by the experience gained at Port Leopold in 1848, Captain M'Clure determined to get his parties away at once, instead of waiting, as Sir James Ross had done, until May 15th. The sledges were therefore laden; and although, with provisions for six weeks and their equipments, every sledge weighed eleven hundredweight, and there were only six men to drag it, they moved, on trial, at a rate which gave good promise of successful journeys.

Each of the three sledges was to take a separate course: one, commanded by Lieutenant Haswell, was directed to proceed to the S.E., following the coast of Prince Albert Land, towards the land known to exist north of Dolphin and Union Straits, and named by its discoverer Wollaston Land; another sledge, under Lieutenant S. Gurney Cresswell, was to follow the coast of Baring or Banks Land, to the N.W.; whilst the remaining party, with Mr Wynniatt (mate), was charged with the duty of examining the coast of Albert Land to the N.E. towards Cape Walker.

On 18th April 1851, the sledges of the Investigator left the ship* with the hearty good wishes of all on board; and, like their brother seamen of the expedition then wintered under Griffith Island, they held on their toilsome course in spite of cold, hardship, and every difficulty, cheered by the then still strong hope of finding Franklin's lost expedition. To follow each party in its arduous and monotonous labours, would be but an uninteresting repetition of an oft-told tale; yet the general reader should be reminded how nobly those gallant seamen toiled who were despatched from the Investigator, or from other ships, to search on foot for our missing comrades. Sailors by profession, and consequently unaccustomed to long marches or to dragging heavy weights—the major portion of their lives having probably been

* The sledges of Captain Austin's expedition, then wintering at Griffith Island, left, it will be remembered, three days earlier—namely, on 15th April 1851.

spent under a broiling sun on the coast of Africa, or in the East or West Indies—we yet see these men readily enter the arctic regions, and push into the sea beyond the boundaries of our knowledge of the earth's geography, and even of the limits of the wanderings of the hardy Esquimaux.

Imprisoned as they had been for a long and dark winter, entirely left to their own resources for health, food, and amusement, rationed upon the coarse and endless repetition of salt beef and salt pork, varied with occasional preserved meat to check the slow but certain march of scurvy, they were now sent to travel upon snow and ice, each with 200 pounds to drag—an inevitable load, for it comprised food, fuel, raiment, sledge, and tent. If they should feel cold, they must be patient; for until they return to the ship there will be no fire to warm them. Should their parched tongues cleave to their mouths, they must swallow snow to allay their thirst, for water there is none. Should their health fail, pity is all that their comrades can give them, for the sledge must move on its daily march. If hungry, they must console themselves by looking forward to being better fed when the travelling is over, for the rations are necessarily, in sledge-journeys, weighed off to an ounce; in short, from the time they leave the ship until their return to it, the service is ever one of suffering and privation which call for the utmost endurance and most zealous energy.

Severely did the spring of 1851 test the best qualities of the British seaman, not only among the sledge-parties from the Investigator, but also among those of Captain

Austin and Captain Penny; and in every case the result was the same. No man flinched from his work; some of the gallant fellows really died at the drag-rope; others by frost-bites became cripples for life; but not a murmur arose in any party: as the weak fell out from the sledge appointed to the longest and most severe journey, there were always more than enough of volunteers to take their places.

An attempt has been made to decry the labours of these seamen in the search for Franklin, and to compare them with the deeds of the Hudson Bay voyageurs. The comparison cannot be made with justice to either side. The voyageur works from a great continent, forming a sure and safe base of operation—peopled wherever he has to go by Esquimaux or Indians, consequently capable of supporting life with ordinary skill and foresight. The major portion of the search commenced in 1848 from the Hudson Bay Company's Territories has been carried on in boats and canoes; and the wintering places on Peel River and at Fort Confidence being south of the arctic zone, the severe trial of health, occasioned by a three-months' absence of sunlight and salt-meat rations, is happily avoided.

The work the voyageurs were called upon to execute they have done well, and if placed upon the barren lands of 74° to 78° north latitude, they might possibly undergo the privations of that rigorous climate, its months of darkness and years of unwholesome dietary, equally well as our sailors: upon that point we have nothing to say, except that they have not been so tested. It can in no way detract from the high merit of the Hudson Bay servants, in the search for Franklin's expedition, to say that the climate and resources of the shores of the American continent, the scene of their labours, are very superior to those of the sterile latitudes over which our seamen toiled, and the question in no way involves the personal merits of either the men or their leaders.

The fact of the Esquimaux having perished off the face of the region searched by the sledge-parties of Captains Ross, Austin, Penny, Kellett, and Belcher, tells its own tale. English seamen have had to exist, and labour severely, where even the aborigines had found it impossible to live. It is hardly to be expected that in our generation the laurel will be awarded where it is due, but we safely leave to the judgment of posterity the record—and it will bear the closest scrutiny—of how British seamen have laboured in a noble cause. Their reward, poor fellows, has been but small; and living as they do, by the sweat of their brows, shattered health to them brings starvation. They have laboured hard, and deserve well of their country and profession. Had all their leaders been as single-hearted, as upright in purpose, and as stanch, the wretched tale brought home in 1854 by Dr Rae would not have had to be told.*

* The editor does not know of any sledge-journey which can more vividly depict the sufferings which some sledge-parties of sailors went through, than the one of which the following is a brief extract, from the daily journal of the officer in command, the present Captain George H. Richards, an officer second to none in the indomitable energy and skill he has displayed in the successful execution of every duty intrusted him in arctic service.

"On the 22d February 1854," says Captain Richards, "the temperature having ranged between 34° and 45° *minus* for the last four days, I started with two sledges, by Captain Belcher's orders, for

Whilst the sledge-parties of Captain M'Clure's ship, as well as those of Captains Austin's and Penny's expeditions (see 'Stray Leaves from an Arctic Journal'), are plodding along on their arduous search, I must, to con-

Beechey Island, fifty miles distant. After eight miles' dragging, the men were so very tired, cold, and miserable, that they hardly had patience to wait for their frozen meat being thawed; and that eaten, they threw themselves down in their blanket bags, half frozen as they were, to sleep. Next day (the 23d) the thermometer registered 40° below zero, or 72° below freezing-point!" The poor fellows dragged on as well as they could; but the Captain's hands were too cold, and his ideas too much engaged in attending to their safety, to write any journal beyond the hasty but graphic expressions in his note-book,—"It's distressingly cold!" "the pork as brittle as resin;" "the rum frozen!" So fatigued were many of the men, and so debilitated from constant suffering, that their stomachs rejected what food they attempted to swallow. On the 24th, the temperature had fallen to 74° below freezing-point. It seemed as if human endurance could go no further; yet they tugged on, for anything was better than returning to the wretchedness they had left on board their ship. Their noonday meal, called lunch, could not be partaken of; for the rum and the bacon were solid, and they were too cold to wait whilst either thawed. Passing by where the gallant Frenchman Bellôt had fallen a sacrifice in attempting to carry out the orders of Sir Edward Belcher (*vide* Blue-Books), the worn-out and exhausted crews encamped at last off Cape Grinnell.

Another night of sleeplessness passed, for the cold was too intense for the most tired to sleep.

On the 25th February the jaded crews made their way across Griffin Bay, the temperature still so low, and their sufferings so intense, that they could neither eat nor sleep,—a glass of grog and a bit of biscuit being all their food. On the next day the temperature was still 73° below freezing-point (—41°of Fahr.); exhaustion was apparent in all the party, and Captain Richards had, as he says, "serious misgivings as to whether he should be able to proceed." On making the attempt, frost-bites became frequent and threatening; but a fresh gale from the north fortunately blew their sledges on, and in the evening they camped near Point Innes. On the following day Captain Richards and Mr Herbert pushed on to the North Star, at nect the thread of our narrative, remind the reader that we left H.M.S. Enterprise, Captain Collinson, consort of the Investigator, in the Pacific Ocean. She reached the latitude of Icy Cape as late as the 22d of September 1850, having made a long and circuitous passage from the Sandwich Islands. The pack ice was there met; and with winter evidently closing in, the prospect of rounding Cape Barrow that year was at an end. In obedience, therefore, to the strict injunction contained in his orders not to winter in the pack, Captain Collinson bore up for a warmer climate, so as to have his crew and ship ready to resume their labours in the season of 1851.

All Captain Collinson knew of the position of Captain M'Clure, was the report of Captain Moore, of the Plover, who on 5th August 1850 had seen the Investigator, under a press of canvass, steering northward off Wainwright Inlet. Unfortunately one of the many rumours, easily to be traced to the Investigator's communication with the natives of the north coast, which reached the Plover in her idle winter-quarters, induced Captain Collinson to allow an enterprising young officer, Lieutenant Barnard, to be landed in the Russian north-west American settlements, in order to inquire into their truth; and in carrying out this service he was brutally murdered by savages in a surprise of one of the Russian posts, called

Beechey Island, for aid; and once arrived there, both he and his men fervently thanked their God for his protection through no ordinary suffering. It required a week's rest to restore his men to health and strength; and perhaps the most painful part of this tale of suffering is, that it all arose from an idea upon Captain Belcher's part that he was gifted with prophetic powers as to a high range of temperature after the 22d February.

Darabin Redoubt, not far from Norton Sound. The circumstances under which it occurred are related in the following letter of his companion, Mr Adams, assistant-surgeon:—

"Garishka, Russian Fishing Station, Norton Sound,
N.W. Coast of America, 3d March 1851.

"SIR,—The information I have been able to obtain here, appears to be more probable than that which I gained at Michaelowski. It is to the following effect:—

"Soon after Lieutenant Barnard's arrival at Darabin, a Russian and two natives were sent to the Koúkuk river to trade for skins, and they took a letter from Mr Barnard to be forwarded to the Englishmen on the Ekko. These three men were murdered by the Indians.

"On the morning of the 16th of February the governor of the redoubt (Maxemoff or Darabin), who was sleeping in the same room with Mr Barnard and Boskey, hearing a noise outside, went to the door; and immediately on opening it, he was killed by a spear.

"The Indians then rushed into the room; Mr Barnard seized his gun, one barrel of which happened to be loaded with a cartridge, and wounded a man in the arm; he then struck with the butt, until the stock broke; he was severely wounded in the abdomen by a spear, but I cannot learn that he received any wounds from arrows.

"Boskey was badly wounded in the abdomen by two arrows, in the hands by a spear being drawn through them in attempting to wrest it from an Indian, and in the arms by a knife. I can learn nothing of the other Indians, except that they killed one native.

"The inhabitants of the two villages, Tollúkok and Koltargar, were at Oomalartof at the time of the attack, and all were killed,—men, women, and children,—to the number of about fifty. The six who escaped were sleeping in the bath-house at the redoubt.

"I cannot ascertain the number of the attacking party, only that there were 'plenty' of them. Each man carried a shield of thick wood, which was musket-proof; and after the first attack, they appear to have planted them in a line, so as to form a wall, from behind which they fired at the surviving inhabitants.

"There appears to have been no motive for the attack, and so unexpected was it that they were sleeping with their doors unfastened.

"I have seen some of the spears here; they are large, and appear to be of European manufacture; they are inlaid with brass and copper.

"I have added to the enclosed, tracing all the information I have been able to obtain relative to the situation and names of villages and rivers.

"On the 5th of January last Mr Barnard sent a native of this village to the Plover with despatches; he has not been heard of since, and the natives are all so much frightened that I cannot get another to go. I therefore leave these papers with the Russian in charge of this station, to be forwarded if possible.

"We leave this to-morrow.—I have, &c.,

"Edward Adams,
Assistant-Surgeon, R.N.

"Commander T. E. L. Moore,
H.M.S. Plover, Grantley Harbour."

The finale of this sad catastrophe is briefly told, in the handwriting of poor Barnard, in the annexed note to Dr Adams. It speaks volumes for the nerve of the gallant officer, and it is strongly characteristic of the man:—

"DEAR ADAMS,—I am dreadfully wounded in the abdomen; my entrails are hanging out. I do not suppose I shall live long enough to see you. The Cu-ú-chuk Indians made the attack whilst we were in our beds. Boskey is badly wounded, and Darabin dead.

"I think my wound would have been trifling had I had medical advice. I am in great pain. Nearly all the natives of the village are murdered. Set out for this with all haste.

"JOHN BARNARD."

The Russian letter on which this was written bore the date of 5th February, Darabin Redoubt; Russian time being twelve days later than ours. The writing betrayed the agony of the gallant writer, and parts were nearly illegible.

CHAPTER XIII.

SIGNS OF SUMMER INCREASE—SHOOTING-PARTIES ARE SENT OUT—NARROW ESCAPE OF WHITEFIELD—LIEUTENANT CRESSWELL RETURNS, HAVING ASCERTAINED BANKS LAND TO BE AN ISLAND—CURIOUS APPETITE OF A BEAR—LIEUTENANT HASWELL RETURNS, WITH INTELLIGENCE OF ESQUIMAUX BEING AT HAND—CAPTAIN M'CLURE VISITS THE ESQUIMAUX—THE PARTY UNDER MR WYNNIATT RETURN—JUNE IN PRINCE OF WALES STRAIT—A GLANCE AT THE OTHER EXPEDITIONS WINTERING IN ARCTIC SEAS, UNDER AUSTIN, PENNY, ROSS, AND DE HAVEN—A MIDSUMMER SCENE IN PRINCE OF WALES STRAIT—THE POLAR SUMMER'S NIGHT—THE FLOE BREAKS UP—INVESTIGATOR AGAIN FREE—COMPASSES REFUSE TO TRAVERSE—THE SHIP BESET, AND DRIFTING TO THE NORTH-EAST ALONG THE EASTERN SHORE—WOOD-CURRENTS—TIDES—NO PASSAGE FOUND—CAPTAIN M'CLURE DECIDES TO TRY ANOTHER COURSE, BY GOING ROUND BANKS LAND.

Between the 22d of April and 5th of May the signs of approaching summer increased rapidly in the neighbourhood of the Investigator. Every indication of thaw, heat, and vitality was keenly watched and minutely noted; indeed, these observations, and fluctuating hopes and fears for their brother shipmates absent in the sledge-parties, formed the constant round of the existence of those whose good fortune it was not to share in the labours and occupations of sledge-journeys. One day a small lemming was caught, and its fur having changed slightly, from pure white to a faint brown, was a prognostication little likely to disappoint them of the snows being about to melt away from the surface of the smothered land ;—it was the olive-leaf in the mouth of the dove. On another occasion, the quartermaster, whilst clearing the ice off the surface of a hole in the floe, which was always kept open in case of fire occurring, was charmed to see a seal pop his head above water, and stare wonderingly, with his big lustrous eyes, at the blunt Yorkshireman who was intruding upon his dominions.

Some there are who might have spared the poor seal; but the "Man of Hull" hardened his heart, for he thought of the savoury fry it would yield, and straightway poor Poussey * was transfixed with a lance, and his skin, oil, and flesh were soon afterwards contributing their respective quota to the health and comfort of our navigators.

Then a magnificent polar bear, a real giant, ten feet long, with footprints twelve inches in diameter, bore down to survey the Investigator. It was of course fired at, but fortunately escaped with life and skin. The fox and ptarmigan were seen together on Princess-Royal Island and Albert Land: the feathers of the latter lying about in profusion, denoted that they pay dearly for frequenting such distinguished society.

The vessel was now caulked and painted, hatchways opened to dry long-accumulated damp between decks; the holds were re-stowed, after provisions and stores had

* A term borrowed by our whalers from the Greenlanders.

been surveyed; and lastly, a close examination of the crew was made by the surgeon, Dr Armstrong, and its result was most satisfactory. All were in most perfect health. Not a trace of scurvy was detected among the men then on board; "an instance of sanitary wellbeing," as Captain M'Clure justly observes, "unparalleled in the annals of polar voyages." May brought in a temperature ranging from 6° to 30° of Fahrenheit, the wind varying from S.W. to N.W., with occasional falls of snow. On the 6th of May Mr Wynniatt's sledge-party returned to the ship: that officer had broken his chronometer, and wanted to be supplied with another; but there was not a spare pocket-instrument on board, and Captain M'Clure, pained beyond measure at the loss of time already incurred by the return of this party from a position nine days' journey in advance of the ship, despatched Mr Wynniatt again upon his original route during the course of the day. Mr Wynniatt reported that throughout his journey traces of musk-oxen and deer had been very plentiful and fresh; and as the latter animal had also been seen upon the land abreast of the ship, two shooting-parties were established to endeavour to secure an addition to the resources of the commissariat.

On 7th May a sad accident nearly occurred to a young carpenter named Whitefield, one of a shooting-party on the western shore. A large flock of hares had been seen trooping up a ravine just as a heavy snow-storm set in. The rest of the sportsmen retired to their tent for safety, but Whitefield was tempted to go on. Being missed by the others, the men of the shooting-party started two at a time to look for him, each relief running much risk of losing its way and being smothered in drift; yet nothing could be seen of the lost man. Failing in all their efforts, and fairly at their wits' end, the party, which was in charge of a petty officer, retreated to their tent again, and began to fear the worst, when one of them suddenly exclaimed that he heard "the footsteps of a bear!" All heard the sound for a minute, and then it ceased. The drift was so dense they could see nothing; and to their shouts of "Whitefield!" no answer came. Shortly afterwards, during a lull in the gale, some one happened to look out of the tent; and there, not a yard from the tent, knelt poor Whitefield, stiff and rigid as a corpse, his head thrown back, his eyes fixed, his mouth open and filled with snow; his gun was slung over his shoulder, but his body was fast being buried in a snow-wreath. They pulled him into the tent, restored animation, and then sent for aid to the ship. When the man eventually recovered enough to tell his tale, it was strange indeed. He said that, whilst struggling with the snow-storm, and endeavouring to find his way home, he felt a chill, and then a fit came on, which appeared to have deprived him of his senses to some extent, for he had seen people looking for him—some of them had even passed within a hundred yards of him—yet he could neither call them nor discharge his gun for a signal, and, meantime, the snow had covered him. After a while he regained some strength, and fortunately discovered a track leading to the tent, and had actually almost reached it—indeed, they were his footsteps that the people had heard—when again the fit came on, and he sank down a yard from the tent door, in the attitude of supplication in which he was found in the snow. He was fast becoming rigid, and freezing to death, when, by the mercy of Providence, his shipmates saw him. Among the startling narratives of arctic history, there are few more providential escapes.

The quantity of game, in the shape of hares and ptarmigan, seen in every direction by the different shooting-parties, and recorded in Captain M'Clure's diary for May, was very wonderful in so high a latitude; but the sailors and marines, with one or two exceptions, were but poor sportsmen, the sum total of their contribution to the general stock being in four weeks but 156 ptarmigan and seven hares. Yet one valley visited by them was "literally alive with hares and ptarmigan," and large troops of the former were seen by all parties. Keener appetites, however, in the following years made keener sportsmen, as we shall hereafter see.

May 20.—The last of the winter's snow had disappeared from the western side of the strait, as well as from that blessed with a south-eastern aspect; but the opposite shore, which, geologically speaking, was of the same formation, still wore a winter livery, owing to its facing the north-western part of the heavens; the advantage, here so apparent, of wintering upon a coast or in a harbour which looks to the southward, is a point which should be held well in mind by the polar navigator, if circumstances allow him to choose his winter-quarters. At 8 A.M. on this day Lieutenant Gurney Cresswell's party returned to the ship, after an absence of thirty-two days from the Investigator. Lieutenant

Cresswell had searched 170 miles of the coast of Banks Land, from the ship, in a north and north-west direction. For the first fortnight the weather had been most severe, constant north-west gales, dead in their teeth, sweeping down into Barrow Strait. Frost-bites had been frequent, but only two men became seriously injured; and they, poor fellows, being affected in the feet in both cases, mortification of the extremities threatened, and Lieutenant Cresswell had been obliged to listen to the dictates of humanity, and retreat upon the ship just as the weather was improving, and the trend of the coast of Banks Land to the south convinced him it was an island. By this judicious step, however, the men's lives were saved, one only losing a portion of his feet; but a day or two longer on the sledge would have been fatal to both.

The Lieutenant found the north coast of Banks Land, west of Cape Russell, a precipitous cliff of limestone, varying from 1000 feet to 1500 feet in height; while against their base, ice of an amazing thickness had been forced up, by a great north-west pressure, into lofty ridges. Outside this ridge the sledge had made its way for seventy miles, when the land became low at what appeared the western extreme of Barrow Strait. Looking in that direction from a considerable elevation, nothing like land could be seen, the eye roamed over a vast sea of ice; it was again that "Land of the White Bear" spoken of by the natives of Cape Bathurst.

On 21st May an extraordinary event occurred, which was not until afterwards explained. About 10.30 A.M. a large bear was passing the ship, when Captain M'Clure killed it with a rifle-shot. On examining the stomach, great was the astonishment of all present at the medley it contained. There were raisins, that had not long been swallowed; a few small pieces of tobacco leaf; bits of pork fat cut into cubes, which the ship's cook declared must have been used for making mock-turtle soup, an article often found on board a ship in a preserved form; and, lastly, fragments of sticking-plaster which, from the forms into which they had been cut, must evidently have passed through the hands of a surgeon. Captain M'Clure, ignorant of the positions attained by the other ships that had been despatched from England, surmised that there could only be two ways in which these traces of civilisation in the bear's stomach could be accounted for. Either the bear had come over some floe of ice visited by the Investigator last autumn, or the Enterprise was wintering close at hand. Now we know that the Enterprise was then in China, and it is hardly probable, bearing in mind the rapid crushing and churning of the ice, as described by Captain M'Clure, in the month of November 1850, that any of it should have escaped being rolled over more than once. The field for conjecture would therefore have been a wide one, had he, like ourselves at the present hour, known of the relative positions of Austin's, Penny's, and Rae's expeditions, as well as of the quarter which Franklin's people had reached. In such a case the most probable supposition would have been, that from one or other of them Bruin had made his very extraordinary collection of curiosities. So impressed, indeed, was M'Clure with the idea that the Enterprise must be in the neighbourhood, that he despatched Lieutenant Cresswell along the south-east shore of Banks Land, with a sledge provisioned until 10th June, to seek her.

After that officer had left the Investigator, the bear's secret was revealed, for some sportsmen in search of game picked up a preserved-meat tin, around which there were many footprints of a bear; and upon examining its contents, they found therein articles corresponding with those discovered in the stomach of the animal shot on the 21st instant.

On 24th May, her most gracious Majesty's birthday, the Investigator fired a royal salute where perhaps no salute will ever again be fired, and most certainly none was ever fired before. The thermometer, exposed to the sun, rose to-day to 73°; in the shade it fell to 26°. The first gull was observed on the 27th inst., a sure sign that cracks in the floe had already begun to show themselves. An early season it certainly was, and officers and men longed for the open water that was to lead them, as they hoped, to Lancaster Sound.

On 29th May the first-lieutenant's party was seen approaching, and they reached the ship safely soon afterwards. Lieutenant Haswell had been absent forty-seven days, and during that time he had searched a great extent of coast towards the south-east. He reached the extreme point of his journey on the 14th May, when his position was on the north shore of a deep indentation in Wollaston Land, his latitude and longitude being about 70° 45′ N., 114° W. By a remarkable coincidence, Dr Rae, from his winter-quarters in America, reached on the 24th of May (exactly ten days later) a point on the opposite side of the same inlet, the "*extremes*" of the two travellers being thus only forty miles apart. On his homeward journey, Lieutenant Haswell fell in with native Esquimaux, encamped upon the ice, at a place since named Berkeley Point, forming the southern head of the strait. Unable to converse with them but by signs, he pushed on for the ship; and directly Captain M'Clure heard of natives being so close, he made arrangements for going to communicate with them by the aid of Mr Mierching, the Moravian interpreter. Until that moment no one had the least idea they were wintering so near fellow-creatures.

It is true that every part of the coast about the neighbourhood of Princess-Royal Island abounded in Esquimaux ruins; but they were moss-grown and very ancient, and none of the natives of North America met by the Investigators in the previous year, between the Mackenzie River and Point Parry, spoke of land in the direction of Prince of Wales Strait.

On the 30th Captain M'Clure and Mr Mierching started, and on the 2d June reached the Esquimaux encampment, consisting of five tents with as many men, five women, and a due proportion of children.

Three of the men were absent hunting; the remaining two received the visitors, answering the first salute of Mr Mierching with a cry of, "Oh! we are very much afraid! we are very much afraid!" as they probably were; but assurances of the good intentions of the Englishmen soon dissipated their fears. One of the hunters came in shortly afterwards. He is described as a fine, active, broad-shouldered savage, with bow and quiver slung at his back, a large copper-bladed hunting-knife in his hand, well clothed in seal-skins, and his finely proportioned limbs neatly encased in beautifully-made mocassins and overalls. In fact, his appearance, combined with his confiding, frank, and friendly manner, impressed all the party, and marked him as a favourable specimen of the hardy race which wanders over those frigid regions. Game, these people said, such as the musk-ox and deer, was very plentiful, but extremely wild. They assured Captain M'Clure of the continuity of the coast he was now upon with that of Wollaston and Victoria Land. Esquimaux increased, they said, as you went to the south-east; and of all that portion of the coast visited by them they drew a very correct chart, handling pencil and paper as if they were accustomed to hydrography. Mr Mierching understood them, and they him, perfectly, the dialect spoken by the tribe being the same as that of the Labrador coast. They seemed very simple and honest; and when presented with anything, they appeared incapable of supposing that any one would give them an article without expecting an equivalent. A piece of red cloth having being tied by the Captain round the neck of a girl, she ran to the interpreter to know what was to be given in return for it; and when assured that it was a free gift, she gracefully acknowledged it by a smile, and wished to know "what kind of animal it grew upon."

These Esquimaux said that, until they had seen Lieutenant Haswell's party, they had never cast eyes upon a white man, proving pretty distinctly that the lost expedition never reached so far. Copper of the purest description seemed to be plentiful with them, for all their implements were of that metal; their arrows were tipped with it, and some of the sailors saw a quantity in a rough state in one of the tents. Bidding good-bye to these interesting creatures, and promising them untold wealth in the shape of buttons and arms, Captain M'Clure returned to his ship more than ever convinced that if the Honourable Company under whose uncontrolled authority the northern portion of British America has passed as a "vested right," would, when seeking for furs and profits, take a somewhat more enlarged view of their position, and study a little the Christian ruler's duty towards such of their fellow-creatures as it has pleased God to place them in authority over, it would be better for them on that day of reckoning when the support of the great ones of this earth shall have as little weight as large dividends, or stock at a premium.

The sympathy awakened in the mind of Captain M'Clure for the lonely, expiring race of Esquimaux was naturally increased by the isolated position of his ship and crew at this period.*

June 4.—The ship was now surrounded by water, yet the ice of the strait was still seven feet in thickness. The 7th of the month brought back the sledge-party under Mr Wynniatt; his turning-point was on the 26th May, at which time he was only fifty or sixty miles from the farthest point reached by a party under Lieutenant Osborn from Griffith Island. In both cases the land where each party turned back was strikingly similar—

* Captain Collinson, with his ship the Enterprise, wintered in the following year, 1851-52, amongst these people, and their intercourse was most friendly.

low, with off-lying shoals, and closely beset with stupendous ice. Since then, in the winter of 1853-54, two of her Majesty's ships, the Resolute and Intrepid, were caught in the pack, and wintered due north of this intervening fifty miles of ground; and although the wind blew fresh from the north and north-west, they did not drift through any channel in a southern direction.

June is passing slowly, for the water does not make half fast enough for men tired of eight months' imprisonment; but the ice has diminished 2 feet 10 inches in thickness in thirty days, and the water-pools upon the surface are extending towards one another, and boring holes through the floe beneath in all directions. The glistening hummocks are turning to a faded-yellow colour, and silently toppling to decay; the ducks and geese and swans fly cackling by, wondering, perhaps, whether the Investigator is an island on which it might be prudent to deposit their eggs, so as to secure them from the sly Renard who is eyeing them with a watering mouth; whilst the long-silent ravines burst out with a view halloo! and send glacier, snow, water, land, and stone, flying far over the floe which fringes either shore.

But whilst the Investigator is waiting for that mysterious but certain motion of the ice-fields in the strait which will release them, let us cast a glance at the points reached during this spring by the many parties pushed out from Griffith Island and from Captain Penny's wintering-place in Barrow Strait. Thanks to the close attention paid to the details of sledging by Lieutenant M'Clintock* when serving under Sir James Ross, in

* The present Captain Sir Leopold F. M'Clintock, R.N.

1848, and to the vast improvement his ingenuity enabled him to effect in it, the sledge-parties from Captain Austin's squadron were the most perfectly appointed that ever perhaps left on arctic service.* Aided by this splendid equipment,—which only required that more of the officers should have been as skilled as M'Clintock in turning them to advantage to have yielded still better results,—the sledges from the Resolute, Assistance, Pioneer, and Intrepid did an immense amount of work. That which carried off the palm in distance and value of service performed, was led by Lieutenant M'Clintock in person to Melville Island; and about the same time that Lieutenant Cresswell was standing on the north extreme of Banks Land, Lieutenant M'Clintock was on the southern promontory of Melville Island, only fifty or sixty miles from him.

On the south shore of Barrow Strait, Captain Ommanney of the Assistance was leading a party westward. At Peel Sound he detached Lieutenant Browne with two sledges, to examine that channel downwards; while from the American coast Dr Rae afterwards came up it in a boat, and they approached each other until their *extremes* were only 180 miles apart.

It is too late to regret it now; but had the whole

* The editor, who has since served in a squadron where an attempt was made to claim originality upon the head of equipment, feels it but justice to say, that every part of the sledge-scheme carried out by Sir Edward Belcher's expedition in 1853 was grounded entirely upon Lieutenant M'Clintock's original ideas. Here and there, though his suggestions were made use of, something was done to give an appearance of originality, but it was an appearance only; to Lieutenant M'Clintock belonged the merit where there was any.

strength of that division of sledges been turned upon Peel Strait, we should then have reached King William Land, and saved Franklin's crew. There was then a *furor* for Melville Island and for Banks Land; and Captain Ommanney and his adviser, the writer of these pages, got small thanks for their forethought in heeding Peel Strait at all. Captain Ommanney, anxious to pass nothing, went down another opening, fearing it might be a strait; and Lieutenant Sherard Osborn pushed on with another sledge, as far as his provisions would allow him, turning back, as I have said, when about forty or fifty miles from the point reached by Captain M'Clure's party.

Another sledge-party, under Lieutenant Aldrich, searched far up to the northward by way of Byam Martin Channel, whilst Penny, with boat and sledge, opened up and examined Wellington Channel. Not a fresh vestige did these parties discover of Franklin's whereabouts, beyond the fact that his first winter-quarters had undoubtedly been in Beechey Island: this first and most important fact Captain Penny ascertained.

Before returning to the Investigator, a glance must be cast in the direction of Behring Strait. There we see the Enterprise, Captain R. Collinson, C.B., pushing into the ice. On the 29th July 1851, she rounded Point Barrow with some difficulty, and then, following the American shore on the footsteps of the Investigator's route in 1850, we will leave her struggling with and mastering successfully the many dangers of that shoal coast and ponderous ice.

The month of July is the summer season of latitude

70° north. The Investigator has bent sails, hoisted up her boats, and keen eyes from the mast-head watch the daily increase of water which is detaching the floe from either shore. The russet tints of the land on both shores have replaced the tiresome white of winter; the ravines are again silent, the *débâcle* has passed, and the waters only run now in modest trickling streams. Here and there along the edge of some deep cleft in the land the white streak of a pigmy glacier gleams, for the summer heats cannot penetrate there; but on the sunny slopes, or in sheltered valleys, the modest flora of the north spreads her short-lived store—lichens and moss—in rich profusion of species and colour. The lovely golden hue of the anemone and poppy, the purple-blossomed saxifrage, and white flowerets of the London-pride, appear interlaced with the rich green of the ground-willow, and rose-tinted leaves of sorrel: all relieve the wanderer's eye, and carry him back with softened feelings to some nook in his own dear land, where the flowers and trees and herbs, though far surpassing in loveliness those before him, were yet not half so much appreciated.

The plover, phalarope, and bunting, here rear their young, untroubled by man: around the margin of the petty lakes formed by melted snow upon the terraces, wild-fowl of many sorts—the king and common eider, the pintail duck, and the Brent goose—form their simple nests, in spite of the prowling fox and piratical boatswain-bird (*Stercorarius parasitica*), the former in quest of the parent, and the latter of her eggs. And then along the face of some beetling cliff which, fronting to the south, gives good promise of having water early at its base, clouds of shrieking gulls, kittiwakes, and burgomasters hold a noisy parliament. There was no night to overshadow this scene: the sun rose high during the day along the southern half of the heavens, and sloped towards the north until midnight, without setting. There was no darkness now, as during winter there had been no light.

Yet it must not be supposed that, in the arctic regions, there is not a perceptible division of the day into that portion of it intended for labour and that for rest. Between the hours of eight in the evening and four in the morning, in spite of the sun sweeping through the heavens, there is a perceptible change; the light is more subdued, the tints of land and sea less strong, shadows less marked, the birds go as naturally to roost as if it was dark, and nature is evidently reposing. Nothing can be more lovely than this polar night, which is not night, or, as it has been beautifully described, "the long mild twilight, which, like a silver clasp, unites to-day with yesterday, when morning and evening sit together, hand in hand, beneath the starless sky of midnight." They who have once looked at such a landscape can never forget it; and though perhaps the penalties attached to a visit to such scenes may serve to check enthusiasm upon the subject of their attractions, yet those least susceptible to the impressions of the wonderful and beautiful, must, when standing among the marvels of those distant regions, have felt emphatically the truth of those eloquent sentences in our Bible in which the creation is described, and every phase of it declared to be good and perfect.

There was great joy on board the Investigator from the 10th to the 14th of July. The floe had commenced moving and breaking up: a lane of water was seen extending on the former day along the western shore to the northward. The ice in which the ship was still imprisoned on the eastern side drifted about a couple of miles, and then suddenly broke up, leaving her again free, after being fixed to one spot for nearly ten months. The dangers of the navigation now recommenced: the ice was still very plentiful, and the clear water in very small patches; and as the pack drifted to and fro, all the helpless ship could do was to fasten to the strongest masses, and trust to their strength for safety from other fields of ice.

The set of the currents or tides had long been an anxious question with Captain M'Clure. The tide-pole in thirty fathoms water was not a sure guide; but, so far as its help and twelve months' observation enabled him to form an idea, the flood-tide came from the south up the strait, the rise and fall being about three feet at spring-tides, and little, if anything, at the neaps. The prevailing current, judging from drift-wood and other symptoms, was north-east along the eastern coast into Barrow Strait, and on the opposite, or Banks Island shore, if anything, the current set south-westerly. Tempted by the appearance of some clear water upon the western coast of the strait, the Investigator, on the 17th July, cast off, and attempted to reach it. She was, however, caught by the pack-ice, and in a dense fog drifted with the crushing floes so close to Princess-Royal Island as to hear the screams of the sea-fowl on the cliff; and, as in the previous autumn, she only escaped destruction by what seemed a miracle. After many a hairbreadth escape from shoals and nips in the ice, Captain M'Clure decided upon returning again to the eastern coast, and following it, God willing, into Barrow Strait. On the 24th the first object was accomplished by crossing the strait and reaching a spot named Cape Armstrong. Here such a quantity of drift-wood was seen upon the beach that a cutter was sent to embark a load. It was all American pine, some of it so fresh, that the carpenter was of opinion that it could not have been drifted from its native forest—either upon the banks of the Mackenzie or Coppermine River—more than two years since.

A serious and alarming difficulty now added to the anxieties of our navigators. The compasses, without any apparent cause, became exceedingly sluggish, and varied to such an extent in the dense fogs then prevailing, that it became impossible to tell which way they were going. The standard compass one day showed the ship's head to be N., whilst the starboard one pointed S.W. by W. ½ W., and the port compass remained obstinately at S. by W. Every care was taken to ascertain and remove the cause of this eccentricity in the needles, but in vain. The Investigator was again beset in the ice, and with slight intermission continued so until the 15th of August, during which time she drifted about two miles *per diem* to the north-east with it, and eventually reached 73° 43′ 43″ N. latitude, and longitude 115° 32′ 30″ W., in which position she remained at the tantalising distance of *twenty-five miles from the waters of Barrow Strait!*

Further than that no effort could advance the ship,

and there were occasional sets of the ice to the south-west, with N.E. winds, which threatened to send them back from whence they came. The young ice at nights had already begun to form again, the sun again set, and darkness had commenced; Captain M'Clure knew that his days of navigation were already numbered. If he could push into the pack of Barrow Strait, with a prospect of drifting with it to the eastward for Lancaster Sound, he was prepared to do so; but it would be folly merely to get entangled in it at the entrance of Prince of Wales Strait, and be swept back again to winter in 1851-52, in the same position he had passed the previous season. Impressed with this feeling, it was with no small anxiety that, on the 16th of August, he proceeded to take a careful survey of the ice ahead, before he decided upon launching into it, or adopting some other course by which to endeavour to carry his ship through the north-west passage in safety; and to perfect upon one line at any rate the search for Franklin's expedition. He says, "I observed the ice closely packed, extending across from one side of the strait to the other:" it formed an unbroken line without a prospect of successful passage through it for a sailing ship; and then he immediately determined, with that decision which formed the secret of his wonderful success, to bear up, go round the south end of Banks Land, and endeavour, by passing to the westward of it, to reach Melville Island by that route.

CHAPTER XIV.

THE INVESTIGATOR BEARS UP, AND GOES ROUND THE SOUTH END OF BANKS LAND—RAPID PROGRESS UP THE WESTERN COAST—THE LANE OF WATER DIMINISHES—PERILOUS PASSAGE BETWEEN THE NORTH-WEST COAST AND PONDEROUS PACKED ICE—EXTRAORDINARY ACCIDENTS AND WONDERFUL PRESERVATION—NORTH-WEST EXTREME OF BANKS LAND—NO GLACIERS OR ICEBERGS WEST OF LANCASTER SOUND—DISCOVERY OF ANCIENT FORESTS—ARCTIC LAKES—FRESH-WATER FISH—THE INVESTIGATOR DRIFTS INTO THE PACK IN AN AUTUMNAL GALE—ESCAPES AND STRUGGLES ALONG-SHORE—SEPTEMBER NIGHT-SCENE OFF BANKS LAND—23D SEPTEMBER 1851, RUN ASHORE DURING THE NIGHT IN THE BAY OF MERCY—SHIP AFLOAT—FAIL TO GET INTO THE PACK OF BARROW STRAIT—WINTER-QUARTERS, 1851-52—REDUCTION OF ALLOWANCE OF FOOD—LAND FOUND TO ABOUND IN GAME—WANT OF GOOD HUNTERS—ACUTE INSTINCT OF THE REINDEER—ARCTIC HARE, WOLF, AND FOX—CONTINUED GOOD HEALTH OF THE CREW—CLEVERNESS OF THE ARCTIC RAVEN—THE POLAR BEAR AND ITS HABITS—VIOLENCE OF WINTER SNOW-STORMS—CHRISTMAS-DAY—THE ARRIVAL OF H.M.S. ENTERPRISE IN PRINCE OF WALES STRAIT—SHE FAILS IN ROUNDING BANKS LAND, AND WINTERS AT THE ESQUIMAUX SETTLEMENT IN WALKER BAY.

THE helm of the Investigator was "put up." The good ship which had so gallantly striven to escape through the northern outlet of the strait, by which the existence of a water-communication between the Pacific and Atlantic Oceans had been discovered, wore round upon her keel, set all sail, and sped rapidly to the south-west, passing Princess-Royal Island for the last time. The officers and crew were astonished to find that not a particle of ice was to be met; floes, hummocks, huge piles of ice that had fringed the coast so recently, had all disappeared! After a run of 100 miles in clear water, the 17th August found them passing the majestic cliffs of Nelson Head, the southern extreme of Banks Land, the land preserving the same bold features for 25 miles more to the westward, where it terminated at Cape Hamilton. Here they encountered a heavy swell from the S.E. with a fine breeze, which made the Investigator throw up her heels, much to the delight of those who were on board. Gradually turning to the north-west, and then north, Banks Land was found again to resume, in some measure, the same undulating features and long sloping beach that characterised its eastern coast. Considerable quantities of drift-wood lay on the beaches north of Cape Hamilton; much vegetation was seen, and numerous flocks of wild swans and geese were feeding along the shore.

On the 18th of August the Investigator had run the extraordinary distance, in such a latitude, of 300 miles, without being once checked by ice. The pack on this day was seen hanging in a heavy body in the south-west, leaving, however, a lane of six miles of clear water between it and the shore. In the afternoon Cape Kellett was rounded with some little difficulty, the ship passing, with sufficient water to float her, between the edge of grounded ice and the coast. The land was now so low that the hand lead-line became for a while their best guide. The soundings happily were regular, and, aided by it and a fair wind, they advanced apace to the northward. Throughout the 19th the ship sometimes ran as much as seven knots per hour, the width of the lane of water in which they were sailing varying from three to five miles. Noon that day found them in 73° 55′ north latitude, and 123° 52′ 30″ west longitude; and already did Captain M'Clure count upon extending his voyage to the north of Melville Island, and then striking for some strait or sound leading into Baffin Bay!

That night, however, a sudden and remarkable change took place. They had just crossed Burnet Bay, within Norway and Robilliard Islands, when the coast suddenly became as abrupt and precipitous as a wall; the water was very deep, sixty fathoms by the lead-line within 400 yards of the face of the cliffs, and fifteen fathoms water where it actually touched them. The lane of water had diminished to 200 yards in width where broadest; and even that space was much hampered by loose pieces of ice aground or adrift. In some places the channel was so narrow that the quarter-boats had to be topped up to prevent them touching the cliffs upon the one hand, or the lofty ice upon the other; and so perfectly were they running the gauntlet, that on many occasions the ship could not "round-to" for want of space. Their position was full of peril, yet they could but push on; to attempt to retreat was now out of the question. The pack was of the same fearful description as that they had encountered in the offing of the Mackenzie River, during the previous autumn. The surface of the floes resembled rolling hills, some of them 100 feet from base to summit; and the edge of this wonderful oceanic ice rose in places from the water as high as the Investigator's lower yards. Any attempt to force the frail ship against such a pack would have been mere folly; all they could do was to watch for every opening, trust in the mercy of God, and push ahead in the execution of their duty.

If this ice had at any time set in with its vast weight against the sheer cliff, nothing, they all felt, could have saved them; and nothing in the long tale of arctic research is finer than the cool and resolute way in which all, from the captain to the youngest seaman of the gallant Investigator, fought inch by inch to make their way round this frightful coast.

Enough has been said to give a correct idea of the peril incurred at this stage of the voyage, without entering into minute details of the hairbreadth escapes hourly taking place, but one instance may be given as a sample of the rest. After the 20th of August, the Investigator lay helplessly beset off the north-west of Banks Land, for the wind had pressed in the ice, and the ship was almost cradled in floe-pieces. On the 29th of August, however, a sudden move took place, and a moving floe struck a huge mass to which the ship had been secured; and, to the horror of those on board, such was the enormous power exerted, that the mass slowly reared itself on its edge, close to the ship's bows, until the upper part was higher than the fore-yard, and every moment appeared likely to be the Investigator's last, for the ice had but to topple over to sink her and her crew under its weight. At the critical moment there was a shout of joy, for the mass, after oscillating fearfully, broke up, rolled back to its original position, and they were saved! Hardly, however, was this danger past than a fresh one threatened, for the hummock to which the ship was secured was impelled forward by the whole weight of the driving pack towards a low point of land, on which, with frightful pressure, the great floes were breaking up, and piling themselves tier upon tier. The Investigator had no power of escape; but every hawser was put in requisition, and hands stationed by them. An attempt to blow up a grounded floe upon which the ship was driving, only partially succeeded; the nip came on, the poor ship groaned, and every plank and timber quivered from stem to stern in this trial of strength between her and the ice. "Our fate seemed sealed," says Captain M'Clure, and he made up his mind to let go all hawsers. The order was given, and with it the wreck of the Investigator seemed certain; all the leader hoped for was, to use his own words, "that we might have the ship thrown up sufficiently to serve as an asylum for the winter." If she should sink between the two contending floes, the destruction of every soul was inevitable.

But at the very moment when the order to "let go all hawsers" was given, and even before it could be obeyed, a merciful Providence caused the berg which most threatened them to break up, and the Investigator was once more saved; though still so tightly was she beset that there was not room to drop a lead-line down round the vessel, and the copper upon her bottom was hanging in shreds, or rolled up like brown paper.

The reader must not confound the idea he may have formed of this wonderfully heavy oceanic ice in the great water space between Banks Land and Behring Strait, with icebergs.

Floe-ice, however stupendous in its proportions, cannot become berg-ice. The former is a salt-water production; the latter is of the land, and launched from the glaciers — which are only to be found in particular localities—and is entirely formed of fresh water or snow.

No glaciers of any size exist westward of the 85th meridian of W. long. in Lancaster Sound, and no true icebergs are met with in the great arctic American archipelago. Here and there in dark and sunless ravines of North Devon and Melville Island a pigmy glacier may be found; but it never reaches the sea, nor forms those great barriers of fresh-water ice which, under the various terms of glacier or *mer-de-glace*, roll on their mysterious and wonderful course from the interior of Greenland, and launch themselves as icebergs into the deep waters of Baffin Bay.

The ice met with by Captain M'Clure was aged sea-ice, which, for aught we know, may be centuries old; and it seems, from the want of outlets, likely to increase yet in thickness to an unlimited degree. The accumulated action of repeated thaws, and the almost constant fall of snow upon the upper surface, have given a peculiar hill-and-dale appearance to it, and rendered much of the upper ice as fresh to the taste as if it had been formed on shore instead of upon the sea.

The heavy westerly gales force a portion of this prodigious ice through Melville Strait; but, either from decay or other causes, it is never seen in its pristine proportions east of Griffith Island and Cape Walker: a floe fifteen feet thick in Barrow Strait or Wellington Channel was a very great rarity, from four to eight feet being the average size.

The Investigator remained some time beset upon the north-west extreme of Banks Land; and whilst detained in that dangerous locality her officers and men rambled into the interior, and they found it far from so sterile as the prospect from the sea had led them to anticipate. Traces of musk-oxen and deer abounded, and both these animals were seen; but the most extraordinary discovery of all was a great accumulation of fossil trees, as well as fragments not fossilised, lying over the whole extent of the land, from an elevation of 300 feet above the sea to its immediate level. Writing on the 27th of August, Captain M'Clure, speaking of this wood, says, "I walked to-day a short distance into the interior: the snow that had fallen last night lay unthawed upon the high grounds, rendering the prospect most cheerless. The hills are very remarkable, many of them peaked, and standing isolated from each other by precipitous gorges. The summits of these hills are about 300 feet high, and nothing can be more wildly picturesque than the gorges which lie between them. *From the summit of these singularly-formed hills to their base abundance of wood is to be found; and in many places layers of trees are visible, some protruding twelve or fourteen feet, and so firm that several people may jump on them without their breaking. The largest trunk yet found measured one foot seven inches in diameter.*"

Again, on September 5, some miles from the hills

just alluded to, Captain M'Clure says:—"*I entered a ravine some miles inland, and found the north side of it, for a depth of forty feet from the surface, composed of one mass of wood similar to what I had before seen. The whole depth of the ravine was about 200 feet. The ground around the wood or trees was formed of sand and shingle; some of the wood was petrified, the remainder very rotten, and worthless even for burning.*"

At a subsequent period to that we are speaking of, a similar kind of fossil forest was discovered nearly 120 miles farther north.

In the spring of 1853, one of the sledge-parties under Lieutenant Mecham, of H.M.S. Resolute, was travelling across the newly-discovered island called after H.R.H. Prince Patrick. When in 76° 15′ N. lat. and 121° 40′ W. long. he says:—

"*Tuesday, May* 31.—Discovered buried in the east bank of the ravine, and protruding about eight feet, a tree of considerable size. During the afternoon I found several others of a similar kind: circumference of first and second tree seen, three feet; of another, two feet ten inches. From the perfect state of the bark, and the position of the trees so far from the sea, there can be but little doubt that they grew originally in this country. I sawed one through: it appeared very close-grained, and was so immensely heavy that we could carry but little of it away."

Geologists and botanists must decide how it comes to pass that trees in the perfect state Lieutenant Mecham discovered them are to be found in such localities. Sir Roderick I. Murchison, an eminent authority, has kindly favoured the writer with some valuable remarks upon this subject, as well as a general geological sketch of this arctic archipelago. It will be found in those remarks * that Sir Roderick Murchison has good grounds for being of opinion that all this timber was floated from a more southern continent at a period when Banks Land and Prince Patrick Land were submerged. A very different climate must then have existed in those regions to allow drift-wood so perfect as to retain its bark to reach such great distances; and I may be perhaps allowed to remark, that when the polar sea was sufficiently clear of ice to allow such timber to drift unscathed to Prince Patrick Land, might not fir-trees have then grown in a soil naturally fertile?

In any point of view the field for conjecture is a wide one, and the navigator and traveller may be pardoned for expressing astonishment when, amidst the wastes of those frozen lands, he finds the wreck of ancient forests where the ground-willow and dwarf birch can now hardly exist.

On the 1st September winter appeared to have overtaken the Investigator in her forlorn position. From the highest land near them the officers and men had in vain looked out over the pack for the hope of release which even a yard of water would afford: all was ice over the surface of that really frozen sea. Keen and strong already came the north-west wind. What would it be in the depth of winter? they asked each other with a shudder. The wild-fowl had nearly all gone south, and the gallant little snow-buntings were mustering to

* See Appendix.

depart likewise. The prospect was not cheering; yet none could complain, for they had come a marvellous distance in the short navigable season of the polar seas, and the distance yet to be accomplished to reach those waters which had been traversed by ships from the direction of Baffin Bay was small indeed compared with that already passed. Squeezed up and cradled in ice about fifty yards off the shore lay the Investigator, and Captain M'Clure expected to have to winter in this exposed position. He therefore commenced to take all necessary measures, though the danger of such a wintering-place was well known to himself. The prospect in nowise improved between the 1st and the 10th of September. The temperature had fallen to +16° Fahr., or 16° of frost; and the aurora borealis flickered its pale light at night through the cold heavens. Everything spoke of winter; yet the position of the ship was too insecure a one to justify the captain in making any final preparations for sheltering the men from its rigour, such as clearing decks and spreading housing, lest some fresh movement in the ice should require the vessel to be again placed under canvass—a wise precaution, which, as we shall see, enabled him to reach secure winter-quarters, and saved his ship.

Amongst other remarkable proofs that the daily excursions of men and officers brought to light, of the land, barren as it was, possessing considerable resources in the shape of animal life, the discovery of lakes with fish in them was not the least worthy of note. Two of these lakes had attracted the attention of the officers, from the extraordinary fact that, although within 100 yards of each other, and possessing exactly the same aspect, one of them was firmly frozen over, while the other had not a particle of ice upon its surface. The only respect in which they were found to differ in relative position was, that the unfrozen lake was ten feet nearer the level of the sea than the other, and its depth was six fathoms, whilst that of the frozen lake was but five—a difference, however, which could hardly account for the fact, the water in both being remarkably pure, and the temperature by thermometer differing only 1°. When Captain M'Clure visited the lakes on the 6th instant he found both frozen over; but the ice in the lower one was only half the thickness of that in the upper; and, to add to the interest attached to this little freak of nature, the lower lake was full of fish—salmon-trout, varying from three inches to a foot in length—whilst the upper one had not a living creature in it. The exquisite transparency of the young fresh-water ice enabled him to ascertain this fact as easily as if he had been looking through a crystal.

Ancient traces of the Innuit or Esquimaux were found here, showing that, even to this remote corner, that extraordinary race of hunters and fishers had at one time extended their wanderings; and not far from the ruins of their huts and *caches* some more hills were discovered, in which there existed a considerable stratum of wood, "with trees," says Captain M'Clure, "of considerable length and diameter projecting from the sides of the hills, and that, too, in a state of preservation which rendered them not unfit for firewood."

On the 10th September the wind veered to the

southward, the temperature rose, and at midnight the ice went off from the coast, without the slightest warning, carrying the poor Investigator with it, and handing her thus over to the tender mercies of the much-dreaded pack-ice. Fortunately the ship was on its weather-edge, although so cradled by ice under her bottom as to be helpless; and painful were the feelings of all on board until their position was ascertained at daylight. But amid the roar of the gale and tossing of the floes which had caused this sudden danger, the firm hand of the leader wrote in his diary:—

"Thus we launch into this formidable frozen sea. *Spes mea in Deo.*"

Daylight showed them to be drifting north-east, one mile off shore, in 100 fathoms of water, at the rate of about a mile an hour. So far it was consolatory, as it was the direction they wished to go; but their great object was to free the ship, and to secure her in some nook in the land, to avoid the pack when it crashed in again upon the shore, which it assuredly would do directly the southerly gale abated in the least. By dint of enormous charges of powder, placed under and amongst the ice which held the Investigator, this release was at last effected, at a time when neither hawsers, saws, nor chisels were found of the slightest avail. Launching once more into her own element, the Investigator struggled on during the day, and, as night closed in, sought shelter amongst the grounded ice. Another night and a day of continued danger and anxiety followed, for the wind slackened, and the pack again rolled along the coast, pivoting upon the grounded pieces, and threatening their frail bark as it pulverised masses thirty or forty feet thick, or threw them high up on the beach, or atop of one another. Through the long dark night the sullen grinding of the moving pack, and the loud report made by the ice-fields bursting under the pressure, echoed through the solitude; and as the starlight glimmered over the wild scene to seaward, the men could just detect the pack rearing and rolling over by the alternate reflected lights and shadows.

It was a time to try every nerve; and fervently all prayed for some providential circumstance to place them in a haven of security for the winter. That prayer was at last answered; for, having once more freed the ship from the ice which surrounded her, in order that a lane of water stretching eastward might be turned to advantage, the 19th of September saw the Investigator again progressing along the coast. Fifteen miles were accomplished, and at night the vessel was secured as far as circumstances would admit of. Two whales, the first seen for a very long time, passed them on this day, and appeared to be going westward. Next day, struggling with a succession of difficulties which nothing but the unparalleled gallantry and zeal of every soul in the ship enabled them to surmount, the Investigator reached a headland, since called Cape Austin; and here she was secured again, near a place where the floes had run up a steep slope of the land to the height of seventy feet.

On the 22d this cape was rounded, and the voyagers immediately found the appearance of the ice less formidable, and all breathed afresh at the feeling that they were now fairly in the *waters of Barrow Strait!* Whilst pushing slowly on, two small bays were seen, but so choked up with old ice as to render it impossible for the ship to find shelter there. Some idea may be formed of the narrow strip of water along which the Investigator was now seeking her way, from the fact that on one occasion, as they approached a cape, the lower studding-sail boom had to be "*topped up*" to allow the vessel to pass through a crack (for it was nothing else) between the steep cliffs on the one hand and the floes upon the other.

The 23d of September 1851—the last day of the gallant ship's achievements—came in most promisingly. Water was seen ahead long before day-dawn, sail was set, and she battled on all day to the eastward, making a little southing, as the land trended that way. Hitherto Captain M'Clure had avoided pushing on after dark, since the nights had now become so long; but for many cogent reasons he was induced on this occasion to depart from this rule; and, as the result proved, it was unfortunate in one respect that he did so, for about half-past seven o'clock in the evening the ship ran ashore on a steep bank. The crew strained every nerve to get the vessel off; and after clearing the fore-hold and store-rooms, and laying out a stream-anchor and cable, she floated off during the night.

Next day they found themselves in a large bay, affording good winter-quarters, and perceived that it was impossible to round its north-eastern horn. Under these circumstances, and considering the sufferings and labour his crew had already undergone, Captain M'Clure made up his mind to winter where he was; and, in token of his gratitude to a kind Providence, the bay was appropriately called the Bay of Mercy. It was no empty expression, for every heart in that ship was filled with emotion; and many prayed that in after years, should they be spared to reach their homes, the recollection of the bounty and goodness of Him who had upheld them through such anxieties and dangers might never be effaced from their memories.

Winter came on apace, but the Investigators were ready for it. With slight exceptions, the arrangements were much the same as those of 1850-51; and, to judge from appearances, it seemed that all were quite as well able to defy its rigour as they had been on the first occasion. As a precaution, however, to meet the possible contingency of an escape from the polar regions not being effected in the forthcoming year, the painful but necessary measure was adopted of reducing the allowance of food per diem of the ship's company. Captain M'Clure ordered that officers and men should be placed upon two-thirds of their ordinary rations; but happily at this time it was discovered that the land teemed with deer and hares; and although the want of professional hunters caused the loss of many a fine herd of deer, yet when the winter set in, nine deer and fifty-three hares, besides small birds, had been shot, and their flesh added to the resources of the ship. The interior of the land, so far as the walks of the sportsmen carried them, appeared well fitted to support the hardy animals of that latitude. Broad plains of dwarf willow, reindeer moss, and the coarse grasses of the north, were seen; and the herds of deer and troops of hares which were daily reported seemed perfectly marvellous to those who had hitherto believed that little if any animal life existed so far north. The above-mentioned animals, as well as the ptarmigan, never, in fact, left the neighbourhood of Mercy Bay even in the depth of winter; and it was only the cold and darkness which prevented their being shot at that season. It would take a volume to describe the novel and interesting habits of these animals, as observed by those who sojourned in Mercy Bay.

Pressed by the requirements of such a climate, the instincts of all arctic animals seemed more acute than those of similar creatures placed in more favoured climes. They were watchful and wary to a surprising degree; and as they were protected by the open nature of the country, the sportsmen could not always get even within rifle-distance of the deer, although they probably had never seen a human being before; whilst, strangely enough, these herds appeared to entertain no fear of the half-dozen wolves which always lay round them, ready to cut off a straggler, or pick up a giddy fawn.

The arctic reindeer at this season congregated in large promiscuous herds of bucks, does, and fawns, probably for warmth and protection; and, strangely enough, the hares did so likewise. Some troops of the latter were seen, numbering 150 at least; and the roads made by their march through the snow were beaten as hard as ice. I am not aware that this herding of hares, or the fact of the female bearing six or seven young at a litter, has been before noticed.

Apart from the difficulty of stalking down the deer, the presence of the wolves and foxes was found to be a serious drawback; for if a deer was shot and left on the ground, by the time the sportsman had obtained sufficient aid to transport the meat on board, little beyond the head and shin-bones would be left undevoured; and the robber-wolves, taking care to keep out of gun-shot, would howl dismally, as if mocking the disappointment of the hunter.

As cold and darkness increased, and the absence of the sun rendered it unsafe for the crew to leave the vicinity of the ship, the wolves, pressed by cold and hunger, used to haunt her to a disagreeable extent; and the sad prolonged howl of these gaunt creatures in the long nights added, if possible, to the dismal character of the scene. The Investigators vowed vengeance on brutes which, as they declared, not only behaved in a most unsportsman-like, not to say dishonest, manner, but strove to disturb their slumbers besides.

A great deal of snow fell this autumn, indicative of much moisture in the atmosphere, arising from evaporation from the sea, and proving that a considerable extent of water might still exist amongst the pack, though useless for all navigable purposes. In November the temperature fell to —40°, the lowest, perhaps, ever registered at so early a season; and this augured a still more severe winter than had yet been experienced. The crew were, however, generally in capital health, and actuated by the same fine spirit which had carried them through so many difficulties and endeared them so much to their captain and officers. Two ravens now established themselves as friends of those in Mercy Bay, living mainly by what little scraps the men might have to throw away after meal-times. The ship's dog, however, looked upon these as his especial perquisites, and exhibited considerable energy in maintaining his rights against the ravens, who nevertheless outwitted him in a way which amused every one. Observing that he appeared quite willing to make a mouthful of their own sable persons, they used to throw themselves intentionally in his way just as the mess-tins were being cleaned out on the dirt-heap outside the ship. The dog would immediately run at them, and they would just fly a few yards; the dog then made another run, and again they would appear to escape him but by an inch, and so on, until they had tempted and provoked him to the shore a considerable distance off. Then the ravens would make a direct flight for the ship, and had generally done good execution at the scrap-heap before the mortified-looking dog detected the imposition that had been practised upon him, and rushed back again.

Only an occasional bear was seen, and their footprints were by no means common in this neighbourhood. One bear, however, haunted the bay until fairly chased out of it. He is a noble creature, that polar bear, whether we speak of him by the learned titles of "*Ursus maritimus*," "*Thalassarctos maritimus*," or the sailors' more expressive nomenclature of "Jack Rough!" With all her many wonders, never did Nature create a creature more admirably adapted to the life it has to lead. Half flesh, half fish, the sailor wandering in those frozen regions cannot but be struck with the appearance of latent energy and power its every action attests, as it rolls with lithe and swaggering gait over the rough surface of the frozen sea; or, during summer, haunts the broken and treacherous "pack" in search of its prey. Living and stuffed specimens of the polar bear are too common in our museums and zoological gardens to render it necessary for us to attempt more than to convey an idea of its habits.

When not too fat, the pace of the bear is easy and indolent; but their slowest pace is quite as fast as a man can walk; and when excited their speed is truly wonderful, though very far from being graceful. On level ice, the bears at full speed throw themselves ahead by a violent jerking movement of the powerful fore-paws—an ungainly gallop; but they invariably make for rough ice, and it is there that their strength and activity are best displayed. Amongst packed ice neither man nor dog can compete with them. In the Queen's Channel more than one bear was seen going over broken-up ice, rugged and precipitous as the mind can picture, with a facility truly marvellous, their powerful fore-paws and hind-legs enabling them to keep springing from piece to picee, scaling one fragment and sliding down another with the activity of a huge quadrumana rather than that of a quadruped. It is evidently aware of its superiority in such rough and perilous ground, and is generally found at the edge of the belts of hummocks or broken ice which intersect most ice-fields, or else amongst the frozen packed ice of channels such as those of Barrow and the Queen.

There is, however, another reason for bears frequenting hummocks and packed ice, because it is near such spots water usually first makes its appearance in the summer. The seals are consequently there most numerous, and the inequalities of the floe afford the bears cover in approaching their prey. During summer the colour of the polar bear is of a dingy yellowish hue, closely resembling that of decaying snow or ice. The fur is then thin, and the hair on the soles of their feet is almost entirely rubbed off, as with the other animals of glacial regions; but in the autumn, when the body has recovered from the scanty fare of the previous winter, and a large coating of blubber overlays his carcase to meet the exigencies of another season of starvation, the fur becomes rapidly thicker, the feet are, as the season advances, beautifully incased and feathered with hair, and the animal's colour gradually turns to a very pale straw, which, upon particular points of view, as the light strikes it, looks white or nearly so. The nose and lips are of a jetty black; the eyes vary in colour. Brown is common, but some have been seen with those of a pale grey. Their sense of smell is most acute, facilitated, no doubt, by the peculiar manner in which scent is carried to great distances in the pure and frosty atmosphere of the north.

Bears were seen more than once running down the scent exactly as a dog would do; and the floes about Lowther Island, in 1851, looked as if the bears had quartered them in search of seals, as a pointer would do in a field in England; and the snorting noise made by the brutes as they approach men, indicates how much more they trust to their nose than to their eyes; though both, to their sorrow, often lead them into the clutches of our sportsmen. Bears, however, like mother Eve, might plead curiosity as their ruin.

In Wellington Channel, and elsewhere, the writer has seen a bear, in the far distance, going at the hard swinging pace peculiar to the brute when excited; the head thrown forward; whilst every now and then it would halt, stretch its long ungainly neck as if to inhale a fresh whiff of the distant seal, and then again resume its course, as straight to the prey as an arrow to the target.

It is hardly safe to say what the size of the largest polar bear yet seen may have been. Seamen are naturally prone to the marvellous. The Dutch navigators tell of some bears fifteen feet long! one of which fought a whole ship's company during several hours on the coast of Siberia. An old whaler, whom I asked whether bears were numerous in Spitzbergen, vowed he had seen the floes *black* with them! and that a hundred bears in sight at a time during the sealing season, or early spring, was far from an unusual occurrence! We had better, therefore, leave size and numbers an open question. Bears, however, of nine feet six inches long are not uncommon; and they need be strong to master the large seal of the arctic zone, especially the saddle-back and bladder-nose species.

In the water, although both are able swimmers and divers, the bear can be no match for either of the above description of seal; and it is only upon the floe, or on the slippery pieces of ice which constitute the packed ice, that the bear fairly captures and kills them.

The seal, aware of wherein lies its safety, seldom, if ever, is seen twice its length from the water, whether it be the hole which it has itself made through the ice, or the open sea at the floe-edge; and when basking upon the surface of the fields of ice which float about the Polar seas, nothing can exceed its watchfulness. Its magnificent eyes are so placed as to enable the animal to sweep a great extent of the horizon with the slightest motion of the head; the sense of hearing, which it possesses in a remarkable degree, adds to its security; and, to the uninitiated, the restless vigilance of the seal is particularly striking, — now raising its head to look around, now throwing its head into a position for hearing the slightest motion over the crisp surface of the ice, or gazing and listening *down its hole*—a necessary precaution with so keen a hunter as old Bruin.

It is under such circumstances that the bear exhibits a degree of sagacity and skill in securing its prey, which appears to border rather upon the realms of reason than upon those of mere instinct. Its sight and scent tell it the position of the seal; it throws itself flat upon the ice, and, taking advantage of inequalities invisible to our eyes, gradually nears the prey by a quiet and almost imperceptible movement of the hind-feet. The fore-feet are often placed before its black muzzle, so that nothing but the dingy white of his coat is seen, thus preserving a uniformity of colour with that of the floe. Patiently it nears its prey, who mistakes him for a brother seal, or else, misled by idle curiosity, stays gazing until, with one fatal spring, the bear is upon him. Yet even then the bear does not always secure its feast; and what it is to have succeeded in even griping a seal can only be appreciated by unlucky men like our arctic travellers, who have been hours crawling up, dreaming of delicious seal's fry and overflowing fuel-bags, and seen their prey pop down a hole when within a hundred yards of it. The great muscular power of the seal enables it often to throw itself into the water in spite of the bear's efforts to retain it on the floe; Bruin, however, holds on, for it has almost as good diving powers as the seal, and down they go together. Sometimes the bear comes off a victor, owing to the mortal injuries inflicted upon the seal prior to reaching the water; at another time the bear may be seen rising at some other hole in the floe, or getting upon another loose piece of ice, looking heartily disgusted at his want of success.

The bear dives admirably, and is almost as much at home in the water as upon the ice. If a seal is seen upon a loose floe, the bear will gently slip into the sea, swim with merely the tip of his nose above water,* and, diving under the floe, rise at the very hole which the unhappy seal has looked upon as the only road to safety; and it is this clever expedient of the bear which occasions the seal to watch its hole so narrowly. Even on extensive ice-fields fast to the land, where no hummocks exist to cover the approach of the bear, the seal is not safe, for then the bear slips quietly down a hole, and swims along under the ice until it reaches the one where poor poussey is enjoying the sunshine, and thus takes it at a disadvantage.

It is in the early spring, February and March, after the meagre fare of winter, that the bear's feasting-time occurs. The seal is then bringing forth her young: they are born blind, helpless, and unable for ten days to take to the

* The power of keeping all the body submerged, except the nostrils to breathe through, is possessed by, and is common to, the bear, seal, whale, as well as alligator, hippopotamus, and other amphibious animals.

water; and although the poor mothers leave no precaution unheeded to guard their feeble offspring, it is, nevertheless, a perfect massacre of the innocents, the arctic wolf being suspected, from some facts which came under our observation, to be a participator in the havoc.

The voracity of the bear leads him, however, as a retribution, into sad scrapes. The seal instinctively breeds as close to the open water as possible; the consequence is that the floe, during the early equinoctial gales, often breaks up and drifts away in the form of packed ice; a matter of much indifference to the seal, but to the bear a question of life and death. Numbers of them are doubtless so lost along the whole area of the polar sea, especially between Spitzbergen and Greenland, where both these animals are very numerous. With heavy northerly gales, bears are often set down in such numbers upon Iceland as to be dangerous to the safety of the flocks and herds of those northern islanders, and they have been known to reach the coasts of Norway.

The whalers constantly meet castaway bears at some distance from the land. We have been told by one worthy of credence that he has seen them sixty miles off the land, in Davis Strait, without any ice in sight, and evidently quite exhausted. In this manner, no doubt, nature keeps a check upon the too rapid increase of these brutes; for beyond the possibility of the wolf hunting it in packs, and destroying the cubs, as we believe they do, there is nothing else to keep down the increase of bears. Esquimaux are too scarce, and too ill provided with arms, to be a very formidable check upon them. Wherever seals are plentiful, bears have invariably been found numerous : in Barrow Strait in 1850-51, and in the Queen's Channel in subsequent years, they were seen in great numbers. The Danes say they are plentiful about the northern Greenland settlement of Uppernavik during nine months in the year; and from the united testimony of the natives inhabiting the north-eastern portion of Baffin Bay, and that of Dr Kane, who has lately wintered in Smith Sound, we gather that the bears are very numerous about the *polynias*, or water-holes, formed there by rapid tides.

In the summer months, when the polar bear is in high condition, it can easily be hunted down, lacking, as it then does, activity or wind; but in the depth of winter its voracity and vast strength render it a formidable creature to people unprovided with firearms. Under ordinary circumstances, they always studiously avoided all conflict with our seamen, although the dark days and nights, as well as dense fogs, of those high latitudes, sometimes brought Jack and Bruin much closer together than either party desired, without due notice.

It is folly to talk of the polar bear hybernating: whatever bears may do on the American continent, there is only one arctic navigator who ever saw a bear's-nest! Bears were seen at all points visited by our sailors, at all times and in all temperatures; males or females, and sometimes females with their cubs. In mid-winter, as well as in mid-summer, they evidently haunted spots where tides or currents occasioned either water to constantly exist, or only allowed such a thin coating of ice to form that the seal or walrus might without difficulty break through. This was especially the case in about latitude 78° N., that of Queen's Channel; and a similar state of things appears to exist in Smith Sound.

December 1851 was ushered in, in Mercy Bay, with those tremendous snow-storms which are, perhaps, the most awful visitation of polar regions. All the Investigators could do was to remain shut up in the ship, and wonder what the animals of Banks Land did in a snow-drift which almost tore the housing from its many fastenings. A solid moving body of snow rolled along higher than the topmast heads, and, meeting an impediment in the ship, formed a wreath to windward, and piled rapidly up over her, until the weight of accumulated snow broke down the floe in which she was frozen; the inclination of the ship first one way, then another, and the report made by the cracking of the ice under her bottom, startling those unaccustomed to such accidents. An odd atmospheric condition, which has elsewhere been experienced, was observed by Captain M'Clure in one of these storms, which occurred on the 5th of December. The barometer rose to 30.81, higher than it had before been since leaving England; and the aneroid, graduated to 31.50, stood so high that it could not be registered for four days. In a similar storm, early the following year, the barometer rose above 31 inches.

The second Christmas-day was passed in the ice in a manner to call forth the captain's highest encomiums on his noble ship's company, who behaved, he says, in the most exemplary and satisfactory manner. "After divine service, all went for a short walk until the dinner-hour; from then until bedtime, dancing, skylarking, and singing were kept up on the lower deck with unflagging spirit, good-humour, cheerfulness, and propriety; not a man was inebriated, although, with other additions to the daily fare, amongst which was a pound of the most delicious venison to each person, an extra allowance of grog was issued. Would that the happiness of our little community upon the lower deck of the Investigator could have been witnessed by those anxious for our welfare at home! they would scarcely imagine, otherwise, that the crew of a vessel two years upon her own resources in these ice-bound regions, could create such a scene of enjoyment amidst so many gloomy influences." And as if to countersign this opinion of their chief, several of the petty officers assured him afterwards that, during many years' service in her Majesty's navy, they had never passed a happier Christmas, nor one in which there had been a feeling of more perfect unanimity and good-will,—a feeling shared by every seaman and marine in the ship's company.

The officers dined with Captain M'Clure off a splendid haunch of a Banks Land reindeer, weighing about twenty pounds, with at least two inches of fat on it; and it was pronounced to be most deliciously-flavoured meat. In short, the year 1851 passed away very pleasingly; and in spite of the reduced allowance of food upon which all had been for three months, no one was repining or discontented.

On Christmas-day of the previous winter, it will be remembered that the many ships which left England and America in 1850 to rescue Franklin's expedition were wintering in different parts of the arctic regions; but out of all these the Investigator and the Enterprise now alone remained. At page 150 we left the latter, after having rounded the difficult turning-point of Cape Barrow, progressing eastward along the American shore. As in the Investigator's case, Captain Collinson found the water to make along-shore in a lane whose breadth depended upon the position of the different rivers discharging themselves into the Polar Sea; and in their vicinity destroying the packed ice, or forcing it off to seaward by the strength of their currents.

The Enterprise, when off Cape Parry (the promontory which divides the waters of the Coppermine from those of the Mackenzie River), saw to the northward the southern extremity of Banks Land—the Nelson Head of M'Clure. Steering across for it, Captain Collinson, when under that coast, by a strange combination of circumstances, steered up Prince of Wales Strait, and there on Princess-Royal Island discovered the Investigator's depot, and a cairn containing information up to the 15th June 1851. Passing on after this discovery, the Enterprise, on the 30th August, reached the north end of the strait, but only to be foiled, as the Investigator had been, in any attempt to pass beyond it. Captain Collinson then decided upon taking a course exactly similar to the one pursued by his more fortunate predecessor, and, bearing away, rounded Nelson Head, with the intention of struggling along that western route by which M'Clure had a fortnight earlier successfully carried his ship. On September 3, the captain, little thinking of the Investigator having preceded him in his intended course, was astonished to find on Cape Kellett a record placed there on August the 18th. The ice was now too close in for the Enterprise to push on; and no harbour fit for winter-quarters offering itself so high as latitude 72° 54′ north, Captain Collinson bore up to the eastward, and eventually wintered his ship on the eastern side of the entrance of Prince of Wales Strait, close to the spot where Esquimaux had been found by Lieutenant Haswell, as we have elsewhere related, during his sledge-journey in the spring. From Walker Bay, as their winter-quarters were named, Collinson, after passing his first winter in safety, despatched in the spring his sledge-parties; but, unfortunately for them, the labours of Captain M'Clure's parties, and Dr Rae's exertions, prevented little new ground being reached by any of them, neither did they discover the previous winter-quarters of the Investigator, although, as will hereafter be related, a party which reached Melville Island must at one time have crossed the track of Captain M'Clure's sledge during a trip he made to Winter Harbour. Having thus connected the voyages of the two ships, and shown the relative positions of the only vessels of the searching expeditions left that winter in the polar seas, we must return again to the Bay of Mercy.

CHAPTER XV.

THE NEW YEAR 1852—SATISFACTORY STATE OF THE CREW—DEER OBTAINED DIRECTLY THE LIGHT ADMITTED OF THEIR BEING SEEN—SERGEANT WOON, OF THE ROYAL MARINES, SAVES THE LIFE OF A SHIPMATE—KEEN SPORTSMEN—WOLVES—BOATSWAIN'S ADVENTURE WITH THEM—SPRING—CAPTAIN M'CLURE VISITS WINTER HARBOUR, MELVILLE ISLAND—FINDS NEITHER PROVISIONS NOR VESSEL TO HELP HIM—HIS RETURN—FINDS LARGE QUANTITIES OF VENISON HAD BEEN PROCURED—SCURVY MAKES ITS APPEARANCE—INCREASED NUMBER OF SICK—UNFAVOURABLE WEATHER IN JULY—VENISON EXPENDED—WILD SORREL FOUND IN GREAT QUANTITIES FOR A SHORT PERIOD—10TH AUGUST—WATER SEEN IN BARROW STRAIT—MEASURES TAKEN IN CASE OF BEING ABLE TO ESCAPE—RELAPSE IN THE WEATHER—GLOOMY PROSPECT—AN EARLY WINTER COMMENCES—MEASURES TAKEN TO SAVE SHIP AND CREW, IN THE EVENT OF A SIMILAR SEASON IN 1853—CHEERFUL CONDUCT OF THE CREW—SHORT RATIONS—MODE OF LIVING—BANIAN DAYS AND FESTIVALS—CHRISTMAS AND CONCLUSION OF YEAR 1852.

THE year 1852 came in with a keen and steady cold of from seventy to eighty degrees below the freezing-point of water—a temperature which severely tests the vital energies of man. The weather was still what would be called fine; that is, the wind was light, and fine auroras relieved the darkness in a slight degree. The Investigators met the cold as it should be met, with cheerfulness, energetic exercise, and regularity of habits; New Year's Day consequently found but four trifling cases upon the doctor's sick-list—a satisfactory sanitary state, which continued throughout the spring. Directly the daylight began to increase, and the crew were able to extend their walks, they fell in with reindeer in great numbers. Some of the poor creatures, attracted by curiosity, or pressed by the wolves eternally dogging at their heels, approached the ship for protection, but only, of course, to be fired at. Before the close of January several were shot, and their flesh secured; and, according to the diary of an officer, "the hills in the vicinity of the ship were abounding at that time with deer."

Every encouragement was now given to men and officers who were prepared to undergo the fatigue of sporting for the public weal. One person especially distinguished himself not only as a sportsman, but in the execution of any service requiring unflagging energy and marked intelligence; and this was the non-commissioned officer of royal marines, Sergeant Woon. He did good service everywhere; but no better instance can be given of the metal of which such men as he are made, than what occurred on the 4th of January.

A coloured man serving in the ship, whilst out sporting, wounded a deer, and, after following it awhile, discovered he had lost his way just as a fog came on. The temperature was very low, the man was tired, and the peril of his position caused him to lose his presence of mind and to wander about. By great good fortune Sergeant Woon, who was likewise out in quest of game, joined him; but the poor creature was so beside himself with excitement and horror, that every endeavour to soothe him, by promising to take him safely on board the ship, failed. Fits came on, which left the man quite prostrated in strength. By entreaty and remonstrance the sergeant induced him at last to walk a little; but at 2 P.M., when the glimmering twilight called day of that season was fast closing in, the unfortunate man's energies entirely failed, and he sank upon the ground, bleeding at his mouth and nose, and writhing in convulsions. The sergeant saw that all hope of the man saving himself was at an end; and to leave him where he was, many miles from the ship, was to leave him to certain death: he would have been devoured by the wolves even before the process of freezing to death would have released him from his misery. There was no alternative but to drag him to the ship—no easy matter, when the sergeant dared not part with his gun, and the man was one of the heaviest of a fine ship's company. Sergeant Woon, with heroic resolution, set to his task. Slinging both muskets over his shoulder, he took the man's arms round his own neck, and commenced dragging his half-dead shipmate towards the Investigator. The labour was excessive; and the only relief the sergeant had was, whenever he had dragged the body up one side of a hill, or when he came to a ravine, to lay him down and roll him to the bottom—rather severe treatment for an invalid, but it had the merit of arousing the man somewhat from his lethargy. By eleven o'clock at night the gallant marine had thus conveyed his burden to within a mile of the ship; but ten hours of such toil, amid darkness, cold, and snow, now began to tell upon him. He could drag his burden no longer; and, as a last resource, he implored the unfortunate man to make an effort, and tried to cheer him by pointing to the rockets which the Captain of the Investigator caused to be thrown up as a guide to the missing men. Finding, however, that all his entreaties were replied to by a request "to be left alone to die," the sergeant laid him in a bed of deep snow, and started off for assistance from the ship. Aid was already on its way; and Woon met and conducted two out of three parties to where the man lay, and just in time to save him. He was found with his arms raised and rigid in that position, his eyes open, and his mouth so firmly frozen as to require much force to open it for the purpose of pouring restoratives down his throat, whilst his hands, feet, and face were much frost-bitten. His life was, however, saved; and the courage and devotion displayed by the sergeant need no comment.

On 5th February the sun was seen above the horizon to the southward by those whose anxiety to welcome back its cheerful face induced them to climb the adjacent hills; and in the course of a day or two it gladdened the Bay of Mercy. The sportsmen now became more successful, and a day seldom passed without a deer or hare being shot; and keen must the hunger of those sportsmen have been, for more than one of them, when he shot a deer or hare, refreshed himself by drinking the hot blood, or eating a mouthful or two of the raw meat. They found no ill consequences ensue from the unpleasant food. Now and then a few days' holiday had to be given to the game to prevent it being too much scared—a prudent measure, which always appeared to bring the creatures back to their old feeding-ground. The wolves, encouraged, no doubt, by the feeding they obtained from the wounded animals which escaped the sportsmen, became exceedingly bold; and five of them attempted to cut out an Esquimaux dog that had long been the pet of the Investigator. One of these wolves was a perfect giant, standing nearly four feet high at the shoulder, and leaving a footmark as big as a reindeer's. Many a plot was laid to shoot these wary creatures; but all failed this season, while some of the encounters with them were disagreeably close, and the risk sometimes great. Perhaps one of the strangest adventures was that of Mr Kennedy (boatswain), who, whilst out shooting early in April, broke, at a shot, two out of four legs of a fine buck. Evening coming on, and knowing the animal could not go far, he returned to the ship, and next morning early started to secure his game. Arriving at the place, he was disgusted to find five large wolves and several foxes in possession of the deer. Determined to have his share of the spoil, the boatswain advanced, shouting and calling them by every strong term he could muster, yet afraid to fire his single-barrelled gun at the brutes, for fear of their turning upon him, especially as they appeared inclined to show fight, and made no sign of retreat until he was within four yards of them. Even then only four of them moved off, and sat down a pistol-shot off, howling most dismally. "Pipes" picked up a leg of the deer which had been dismembered, and then grasped one end of the half-picked carcase, whilst a large female wolf tugged against him at the other! The position was, to say the least of it, a disagreeable one; and if the music of the four wolves had brought others of their fraternity to the rescue, the consequences of a struggle between hungry wolves and a no less hungry sailor might have been serious. Fortunately Mr Mierching, the interpreter, who was out shooting likewise on an adjacent hill, had his attention attracted by the howling of the brutes, and came to the rescue. He described the scene as the strangest he had ever seen; and so close were Mr Kennedy and the wolf in their struggle for the meat, that he fancied the animal had actually attacked the boatswain. Seeing more bipeds approaching, all the wolves now decamped, saving their skins as usual, and leaving the poor boatswain only twenty pounds' weight of meat, instead of the 120 pounds which his prize would otherwise have weighed.

The rapid rise of the temperature in April decided Captain M'Clure upon preparing to start for Melville Island with a sledge, in the hope of finding some of Captain Austin's ships, or of ascertaining what depot of provisions had been placed there by them, so as to secure a retreat should the Investigator not escape from her present position.

Everything being in readiness, the sledge, with six men and the Captain, left the Investigator on 11th April 1852. The journey was a trying one, for a relapse in the temperature took place just afterwards, and continued until the 25th. The pack, moreover, offered a sadly rough road. Winter Harbour was reached at an early hour on the 28th of April; and sorely disappointed was Captain M'Clure not to find either vessel or provisions; and the hope of one or the other coming to them hereafter vanished when he read a notice, dated 6th June 1851, left by Lieutenant M'Clintock, on the occasion of his remarkable sledge-journey from Griffith Island. Captain M'Clure congratulated himself that he had at all events discovered this fact before any accident had happened to his ship, for otherwise "I should," says he, "most decidedly, and with the fullest confidence of meeting succour, have pushed for Winter Harbour; and if the Enterprise gets into difficulties, Captain Collinson will, I am fully persuaded, do the same." Shortly after this visit of Captain M'Clure to Winter Harbour, a party from the Enterprise, under Lieutenant Parks, likewise reached Melville Island, and must have crossed the trail of M'Clure, for Mr Parks saw one day, at or near Point Hearne, the marks of a sledge and the footprints of men. That neither party should know of the other being so close, is a strong proof, to be added to the many extant, of the difficulty of meeting one another or discovering traces in those frozen regions.

Disappointed, but not desponding, the leader of the party turned his back upon the old winter-quarters of the gallant Parry, and prepared to lead his men back to their lonely home;—nay, so far was he then from even contemplating a necessity for leaving the Investigator, that on his return-journey across the strait between Melville Island and Banks Land we find the following remark:—"When going towards Melville Island we were much delayed by stupendous polar ice. The whole of this was avoided by crossing well to the eastward, where we met much of last year's ice perfectly level, and occasionally a huge flat floe of older date, but still very good walking:

this appearance of the strait is most propitious, giving every hope of a passage through."

All still promised well on board the Investigator; and the sanitary condition of the crew was reported to be most satisfactory on 11th May, the day of the Captain's return. During his absence the stock of fresh provisions had wonderfully increased by the aid of the sportsmen. No less than twenty head of deer were hanging up round the ship, yielding a *thousand pounds of meat;* and the abundance of food justified an increased issue of rations, which were forthwith ordered to be one pound and a half of venison per man, six days in every fortnight; which, together with six days of preserved meat, left only two salt-meat days in every fourteen. One would have supposed that on such fare, with a dry and comfortable ship to live in, scurvy would be impossible; but, as the sequel will show, the progress of that dire disease became most marked, and though the care of the Captain and the skill of the medical men checked it considerably, still the health of the crew was evidently failing.

Although it may be accounted for in many ways, one fact is incontestable, that on the 15th of May the sick-list had augmented to the unprecedented number of thirteen. May it not be supposed, then, that the moral effect of finding no help at hand told upon these poor fellows? In June we find six men in their beds; and on July 1st Captain M'Clure says, "A more unfavourable report was made by the surgeon to-day, relative to the appearance of the crew at the monthly inspection, than I had hitherto received: evident symptoms of debility, with incipient scurvy, in sixteen of the men."

The long absence of fresh vegetable diet might have predisposed these men to scurvy; but considering the quantity of fresh animal food that had so happily been procured for them throughout the past winter and at this particular time also, it appears likely that the anxious feelings awakened by finding neither provisions nor a vessel at Melville Island, or indeed even a promise of any, had quite as much, if not more, to do in developing the seeds of scurvy, than either the nature of their diet or the slight labour of preparing the ship for sea by ballasting and watering her.

There was no doubt that all were anxious to escape another winter; they felt themselves alone, and having come to save others, it would, unless they reached Baffin Bay in 1852, be a question whether they could save themselves. Each man must have felt how questionable it was whether his physical strength would, after going through another winter, be sufficient to carry him to the Hudson Bay settlements or to one of the Danish posts in Greenland—a long distance, in either case, from the Bay of Mercy, in Banks Land.

But whatever effect these feelings may have had upon their health, there is no question of the firmness with which all were prepared to face the difficulties they might yet have before them. They felt there was no hope but in themselves; but that captain, those officers, and those men were equal to the emergency.

May and June went by without bringing any lively signs of summer; birds, indeed, such as the duck and gull, came, but left the place again, as if from the want of fresh-water pools. The snow at last melted off the exposed parts of the land, and the ice began to thaw, forming large pools of water all over its surface.

July opened very unpromisingly: a heavy northerly wind and snow-storm swept over Mercy Bay, bringing back a painful recollection of the winter that they had hoped was past; and, what was worse, on measuring the floe, it was found to be still increasing in thickness, whereas former experience had led them to expect a diminution of about two feet. The ground became so soft from the snow thawing, and the labour of shooting so great, that the sportsmen could no longer keep up a supply of game; and by the 7th of July all the stock of venison was consumed. It is to be remembered, however, that the resources of Mercy Bay had hitherto yielded the crew of the Investigator three meals of fresh game a-week since October 1851. Hardly had they time to regret the loss of this health-sustaining diet, when the indefatigable Sergeant Woon came on board to report that he had just shot two musk-oxen. The carcasses, when conveyed to the ship, yielded 647 pounds of good meat. These oxen were the first that had been killed in Mercy Bay, and were hailed as a godsend. The sergeant in slaying them had had a narrow escape from the rage of the larger of the two; with his last bullet he had only wounded it, and on its approaching him, he had had to discharge his iron ramrod as a missile into its body in order to save himself.

In the middle of the month the ice-mate reported the floes outside to be in motion. All heard a rumbling noise as if the pack was driving along, and the joy was great until, upon further examination, it was found to be only the *débâcle* from a ravine which was pouring its strong current over the floe. As July drew to its close, the view over the sea from the highest land near was unsatisfactory indeed. Ice, ice on every side! no water, and no water-sky! For a while, however, the ravages of scurvy were mitigated by a quantity of wild sorrel being found, and all hands that could be spared were daily employed collecting it—the sick getting the largest share, and then the men's messes being served in turn. Either eaten raw as a salad, or cooked, it was extremely palatable, for this was the first succulent vegetable the crew had partaken of since leaving the Sandwich Isles two years before.

Hope rose high when, about the 16th August, open water was seen in the straits, and the ice of the bay itself began to be loosened from the shore, though it was still confined by the ice outside. A day or two later, however, the bay opened at the outer end, and the imprisoned navigators saw with delight that a broad lane of water extended along the southern shore for ten miles to the eastward. Their hope of reaching it lay in a strong south wind blowing the ice of the bay and the ship out with it to seaward. As to sawing the whole distance between her and the water, it was impossible; before it could be done winter would be on them. There was a chance of such a wind and such a release, and the top-gallant yards were crossed, sails bent, and the tide-pole taken in. Its registry during ten months gave as a result that the tide rose two feet, and that the highest tide was four tides after the full and change of moon. A beacon was erected in lat. 73° 6′ 48″ N. and long. 118° 15′ W., and in a cylinder attached to a pole was placed a record, telling what the Investigators had done and whither they expected to go, "in the hope," says the leader of the expedition, "that it may meet the eye of some future explorer of these sterile regions, and throw some light upon the fate of those who perhaps may never reach beyond these limits."

The expectation of escape was, however, but short-lived. After the 20th of August the temperature fell, slowly but continually; and when the bay, or that portion of it that had been open, again froze over, all felt that summer was past, and some unforeseen accident could alone save them from wintering again in Mercy Bay. Their summer, poor fellows! had been a most cheerless one; the sun, from the cloudy and misty state of the atmosphere, not having been, with few exceptions, seen since May.

By August 24th the *lead of water had closed!* no water was visible in Barrow Strait; and the Investigators were able to walk in all directions over the bay across the young ice. The land rapidly became covered with snow; the vegetation, such as it was, withered; sorrel could no longer be found; warmer clothing became necessary; and the winter of 1852-3 commenced. "It found us," says M'Clure, "ready to combat its rigours as cheerfully as on previous occasions. We were all thinner than we used to be, for we had been twelve months on two-thirds of our allowance; but we were still in good working condition."

When the first week of September had passed, and the chance of an autumnal gale blowing the ship into the pack was at an end, the leader sat down to weigh the course to be pursued to save his men and his ship. If all remained in the vessel till the year 1853, in the hope the Investigator would carry them home, and (as had happened in 1852) the water should not reach them, all would starve!

On the other hand, it was premature to think of deserting the ship, for she was sound and strong, and both the captain's sense of duty and his pride were enlisted in saving to his country and profession the ship that had been intrusted to his charge, and so gallantly had done her work. He therefore decided to send away, next spring, all but thirty of the healthiest men, and with them to remain by the ship, and run the risk of a fourth winter. Assembling his gallant officers and men on the 8th of September, Captain M'Clure announced to them the state of affairs, and informed them that in April next thirty of them would proceed homewards, divided into two parties, one party retreating by way of North America, up the Mackenzie River; the other proceeding to Cape Spenser, Beechey Island, where Captain Austin's notice, found at Melville Island, led them to expect provisions, and a boat with which to reach Greenland and the Danish settlements. The remaining thirty hands, and the officers in charge of stores, were, if possible, to remain with the captain, and endeavour to save the ship next year; if not so fortunate, they were to spend another winter, and then abandon her in 1854, retreating upon Lancaster Sound to such help as the Admiralty would assuredly send when they knew from their shipmates of their necessity.

This arrangement was cheerfully received by this excellent body of men; never was a country or a profession more worthily represented under trying circumstances than by these gallant sailors; and those who thought they would be the first to go home, were soon heard speculating, with praiseworthy generosity, upon immediately volunteering to come out again in the first ship to the rescue of their messmates, and with light-hearted jocularity promising to bring out a good stock of tobacco-pipes for them—an article which happened to be very scarce in Banks Land, and for which all the ingenuity of the seamen could manufacture no substitute. With men of such a spirit all difficulties and hardships before them vanished, and none repined at what Providence had sent them.

Towards the close of September the stock of rum in the ship was surveyed, and a deficiency found to exist, which obliged the issue to be diminished to half a gill per diem. This was a great loss to the men, and the more so that just then no game could be procured, and they were on a bare two-thirds of the rations; a scale of victualling which, unavoidable as it had been for the past twelve months, was slowly sapping every one's strength. Hunger began now to be felt; and although to his men M'Clure pointed out that their hardships fell far short of those endured by many an honoured arctic expedition, still, when sitting quietly in his cabin, the fact, already more than once represented to him by the surgeon, pressed itself painfully on his mind, that unless aid came in the shape of game, the winter could not be passed on the allowance of food the resources of his ship admitted of, and that all hands would have to abandon the gallant ship in the spring; "but nothing," says he, writing on the 16th October, "but the most urgent necessity will induce me to take such a step."

In October the deer and hares began to return to their winter feeding-grounds in the valleys round Mercy Bay; but seventeen men on the sick-list, and all the duties of housing-in the vessel, throwing up embankments of snow to shield her sides from the bitter gales, and otherwise preparing for that season, left little time and few men to spare for shooting-excursions. The whole game-list for the month showed a return of only two deer and nine hares,—a small quantity amongst so many hungry mouths.

When all the necessary work was finished, and the men had nothing else to do but take exercise and keep themselves and the ship clean, the feeling of hunger and weakness somewhat abated, and the medical report in November showed no increase of disease, except that those of a very nervous temperament became easily excited and unreasonable. November yielded but little game, for darkness was fast increasing; yet the number of deer seen was astonishing, and the wolves harassed the poor creatures until, as in the past winter, they almost fled to the ship for protection. The health of the men appeared to improve somewhat: this favourable change arose, Captain M'Clure thought, from a more contented state of mind than when, in the autumn, the first disappointment at finding no hope of release oppressed all in a greater or less degree. "Hungry," he says, "we all are; but, with a little management, the two-thirds allowance, now that we have nothing to do, keeps us from losing health."

Great, indeed, was the ingenuity displayed in making as much as possible of the daily rations; and food, it is to be feared, was eaten in modes in which the quantity was looked to as the sole recommendation. For instance, the salt meat, instead of being cooked, was just thrown into boiling water to warm it and extract the salt, and then eaten *raw*. On these painful details it is, however, unnecessary to dilate; and a pretty good idea of the scanty fare all were on may be conceived from the following description of the mode in which the gun-room officers lived :—

Their stock was all finished, they were all on ship's allowance like the men, and, like them, adopted the system of each in turn being cook or carver for the mess. The carver's share consisted in getting the *last* portion out of the eight into which the food had to be divided—a method which insured, we need hardly say, the utmost impartiality on the part of the carver, the other members helping themselves to their shares before him. The rations for the day were given out every morning; and each ate it, at his own discretion or inclination, at either breakfast or dinner. They had, in fact, but one meal per diem; for the breakfast, if it deserved the name, consisted of a cup of the weakest cocoa, and a small portion of the small allowance of bread; the rest of the bread, and half a pound of salt meat, containing a good proportion of bone, with just enough preserved vegetable to swear by, constituted the other meal. There was a cup of weak tea in the evening; but few were able to save anything to eat with it.

There were two breaks to this series of banian days during the close of the year; the one was on the 26th October, the anniversary of their discovery of the North-West Passage; the other on Christmas-day. An extra issue of food and some wine were given to commemorate two festivals which all felt were the last that little community would spend together; and those alone who have been similarly placed can appreciate the heartfelt kindness towards each other which hallowed these occasions, and made their humble cheer appear luxurious. It was wonderful, too, to see what care and economy of the stores brought from home, as well as the skill of the sportsmen, were able even in these circumstances to effect. The old-fashioned English plum-pudding was still to be seen on Christmas-day—not a very rich one, may be, but good appetite compensated for what it lacked in that respect. There was "Banks Land vension," "Mercy Bay hare-soup," "ptarmigan pasties," and some musk-ox beef which had hung in the rigging for two years and odd months. The goodwill and determination of all to be merry in spite of adverse circumstances, compensated also for whatever might be wanting. The poets amongst the men composed songs, in which their own hardships were made the subject of many a hearty laugh; painters attempted rude illustrations of past scenes of peril or adventure; the comic actors acted; the sick half forgot their maladies, and the whole company tossed care and anxiety to the winds, and felt thankful for the past and hopeful for the future. Had not both officers and men reason to be proud of their comrades? and was not their chief, the captain of this gallant set of men, justified in saying that nothing was impossible whilst such hearts and hands were ready to carry out his plans for the safety of all and the honour of his country? Full indeed was Captain M'Clure's heart of gratitude, and none knew better than himself where it was due. His own words, written on the close of the year 1852, best express his feelings:—

"The new year is about to commence; not one of my original crew has fallen by disease or accident, and all is more promising than I could have ever hoped for. These and all other mercies are alone due to that all-beneficent Providence who has so wonderfully upheld us in our many trials and difficulties: relying, therefore, on Him, I cannot but feel as the wife of Manoah did, and repeat her exclamation: 'If the Lord were pleased to kill us, He would not have showed us all these mercies.'"

CHAPTER XVI.

THE ENTERPRISE—ILL-SUCCESS OF HER TRAVELLING-PARTIES—LATE SEASON—PASSAGE THROUGH DOLPHIN AND UNION STRAIT—WINTER OF 1852-53 PASSED IN CAMBRIDGE BAY—ESQUIMAUX NUMEROUS—TRACES OF THE MISSING EXPEDITION FOUND—GAME AND FISH ABOUND—UNFORTUNATE CIRCUMSTANCE OF NO SEARCHING-PARTY HAVING VISITED KING WILLIAM LAND—THE BAY OF MERCY—REACTION ON BOARD THE INVESTIGATOR AFTER CHRISTMAS FESTIVITIES—EXCESSIVE COLD—WANT OF FUEL AND CONSEQUENT DAMPNESS—VENISON PLENTIFUL—LARGE SICK-LIST—MARCH '53—THE RETREATING PARTIES NAMED, AND THEIR ROUTES APPOINTED—CAPTAIN M'CLURE'S REASON FOR SENDING AWAY THE SICKLY MEN—WOLVES, THEIR VORACITY AND CUNNING—ANXIETY OF THE SLEDGE-PARTIES AS TO CHANCES OF SAFE RETREAT—RETROSPECTIVE GLANCE AT MEASURES TAKEN IN ENGLAND TO RESCUE THE INVESTIGATOR—MR CRESSWELL'S LETTER TO THE SECRETARY OF THE ADMIRALTY, AND FORTUNATE RESULT—CAPTAINS KELLETT AND M'CLINTOCK ORDERED TO MELVILLE ISLAND—THEY REACH IT IN SEPTEMBER 1852—ACCIDENTAL DISCOVERY OF CAPTAIN M'CLURE'S DESPATCHES ON THE PARRY ROCK—HELP AT HAND FOR THE INVESTIGATOR—APRIL '53 IN MERCY BAY—THE FIRST DEATH—CAPTAIN M'CLURE ADDRESSES HIS MEN TO REMOVE THEIR DESPONDENCY—THE DARK AND BRIGHT SIDES OF THE CLOUD—THE UNEXPECTED ARRIVAL OF LIEUTENANT PIM FROM THE RESOLUTE—THE INVESTIGATORS RESCUED—EXCITEMENT AND HAPPINESS OF THE CREW.

HAVING thus brought the narrative to the close of 1852, it is now necessary to cast a glance back to the spring of the same year, when we left the Investigator's consort, under Captain Collinson, wintering at the southern end of Prince of Wales Strait. All her spring travelling-parties returned unsuccessful from long and arduous journeys, in which some of the men suffered considerably from frost-bites and the other consequences of sledge-work in those high latitudes.

The Enterprise's crew were somewhat refreshed during the summer by procuring a fair supply of game and a considerable quantity of fish from the lakes in Prince Albert Land. It was not, however, until so late as September that the Enterprise appears to have been able to make any progress *eastward* from her wintering-place, a direction Captain Collinson decided upon attempting, with a view to penetrate the unknown space lying between him and Cape Walker in Barrow Strait. A channel which he entered proved eventually to be a gulf, and he then endeavoured to pass by way of Dolphin and Union Strait, and reached, on the 26th of the same month, Cambridge Bay in Wollaston Land; and there he passed the winter of 1852-53, of which we are now writing. In those winter-quarters Esquimaux visited them; and one tribe mustered 200 persons. In their possession was found a piece of iron, which many still believe to have come from the missing ships; and Captain Collinson picked up, moreover, a piece of a doorway or hatch-frame. Knowing what we now do of the point which Franklin's people reached in King William Land, and where they perished, the connection of these fragments with the Erebus and Terror appears to be very probable; but Captain Collinson, being ignorant of those facts, could have no idea of how close his ship was to the spot whereon Dr Rae's informants stated they first saw a portion of Franklin's men; and therefore those fragments told him no more than other traces had done which were previously brought home by Captain Penny from Beechey Island.* The land around the Enterprise abounded in game, and the waters with salmon, for 1100 of the latter were cured for sea-service. We shall here leave H.M.S. Enterprise, premising that those on board of her experienced a very severe winter, and that in the following spring her sledge-parties pushed on to the north-east, passing Rae's farthest point of 1851 by a few miles. But they, like him, although at one time within forty-fives miles of King William Land, did not visit it; had they done so, there is no doubt they would have fallen on the traces of those whom they sought, and very possibly might have found the Erebus and Terror locked up in some such ice-bound harbour as that in which the Investigator was imprisoned. No fault, however, can attach to Collinson that he did not do so; but the sad chapter of unfortunate accidents by which the relief of Franklin's expedition has been rendered unsuccessful, would be incomplete did history fail to point these facts out; and it serves to show that the unavoidable fallibility of the reasoning and wisdom of men have alone occasioned the efforts of England to prove in the end

* The Enterprise wintered on this occasion 120 miles from King William Land, where some forty of Franklin's men were first seen by the Esquimaux, and 200 miles from the Great Fish River, the entrance of which, as has been since ascertained, was reached by a boat from Franklin's lost expedition.

abortive; and it does not prove, as some have argued, the folly of our endeavours to relieve the lost expedition.

When the festivals of Christmas and New Year's Day had passed in the Bay of Mercy, there was not much to make men light-hearted or merry, although they were still determined to look as much as possible on the bright side of things. He who tempers the wind to the shorn lamb watched over them in their trial; and it is remarkable, in reading over a daily journal far too minute for the general reader, to see, throughout this season, the remarkable way in which His bounty supported them. On the one hand we see the resources of the ship gradually failing or being reduced to the slenderest supply—for instance, in the important article of fuel; and, as a consequence, the enervated frames of the men had to stand even a more severe trial than before from cold and damp, and this not only in-board, but also without doors, for the temperature of that winter throughout the arctic regions was unusually severe. From 60° to 65° below zero of Fahr. was registered by the Investigator, as well as other ships elsewhere. Yet this extreme cold, so intense that the very ship seemed to suffer from it, and bolts, trenails, and fastenings were heard to crack under the influence of frost and contraction, forced the deer to approach the ship and the sea-shore so closely as to afford venison weekly throughout this trying season, at the rate of a pound and a half of meat twice in the seven days to every man in the ship.

January passed, giving a mean temperature of—44°, or 76° below freezing-point; and this, be it remembered, was endured by men under-fed, scorbutic, and looking forward to making no ordinary efforts to save their lives in a few short weeks' time.

Occasionally, the idea of Captain Austin's squadron being still in Barrow Strait would occur to the Investigators, and aching eyes would then be cast towards the distant shores of Melville Island, which loomed through the cold twilight, with the hope of some sign of help coming from thence.

In February, Captain M'Clure says, "the excessive cold has been much felt; the lower deck, particularly from about twenty feet before the main hatchway to the gun-room bulkhead, has been very damp and wet; and we cannot afford more coal to dry it up. The sick-list has likewise considerably increased, seldom being below fifteen; four or five of the cases decided scurvy, the others mostly diarrhœa and ague, caused, no doubt, by the moisture between-decks. Our providential supply of fresh meat still continues; four deer fell in the first week, although the sportsmen, on account of the weather, were only able to venture out during two days, and that only for a few hours. Surely this is our manna! I am sure it is considered as such by all on board. The deer appear completely spellbound to this particular locality; for a single mile south or north of our position not one is to be seen, but abreast the ship they are met as we land, and, bad shots as we are, the supply of venison has been regularly 200 lb. per week!"

So much did the dampness of the lower deck increase, that hanging stoves had to be got up between the 7th and 14th of February; and every precaution was taken to check the increase of disease, which had already placed in the doctor's hands one-third of the crew of sixty men. These steps were attended apparently with considerable success; for on 1st March the medical report, all things considered, was more favourable than Captain M'Clure had expected.

On 3d March the travelling parties for England, *viâ* America and Barrow Strait, were told off. They consisted of thirty of the most weakly hands divided into two parties of fifteen men each.

Lieutenant Haswell was to take the one, *viâ* Griffith Island, to Cape Spencer, there embark in the boat which Captain Austin said he had left, and in her attempt to reach Greenland. The other party, under Lieutenant Gurney Cresswell, was to retreat upon the depot formed at Princess-Royal Island in 1851, recruit themselves, and, taking the boat from there, push for the Coppermine River, ascend it to the Hudson Bay Territories, and thence home. Captain M'Clure's reason for thus despatching all his sick and weak hands was, that he felt convinced that these men could not survive another winter after what they had gone through in the past one; and thus he gave them the only chance of saving their lives which it was in his power to afford.

The next thing done was to put these travellers upon full allowance of food, so as to enable them to pick up strength as much as possible. On 28th March the weather had sensibly improved, and the change had beneficially affected all the sickly men; indeed, all but two of them were able on that day to take a little exercise. On the 24th the novel event of a wolf being killed occurred. The brute had gorged himself upon a deer which had been shot, and fell a victim, consequently, to his gourmandism. Strangely enough too, considering the length of time the Investigators had been trying without success to shoot these robbers, another wolf was killed by Mr Court (master) soon after in self-defence. That officer, it seems, was out shooting, when two wolves marked him down, and gradually closed upon him; one keeping in front, and the other behind. After trying all sorts of methods to frighten them off, such as shouting, waving his arms, and running towards them, he found that one of the wolves had closed in on him to within twenty yards. Taking a careful aim, he fired and struck it in the throat, but that did not turn it; and having only a single-barrelled gun, it was as much as he could do to load again and kill the savage beast when it had crawled up to within three yards of his person.

As the travellers were to leave the ship upon the 15th of April, the close of March saw all the many preparations for a sledge-journey well in hand. The officers, though cognisant of the risk and dangers which beset their lines of retreat, wisely hid them from the knowledge of the men. The healthy amongst the sledge crews were consequently sanguine in their hopes of success; but many a poor fellow, whose black and swollen limbs hardly served to carry him about the ship, knew in his heart that, although the journey he was about to take would be his only chance for life, yet it was but a very slender one. Despondency there was not, but there was a deep feeling throughout the ship of their sad and forlorn position, met, however, by a child-like confidence on the part of the crew in the wisdom of their leaders' arrangements, and a perfect faith in the good Providence which had sustained them so far.

Threatening, however, as the future looked for the safety of these gallant men, a series of fortuitous circumstances—providential ones would be the more correct term—was now bringing about their rescue.

It will be remembered that, when Captains Austin, Penny, and Sir John Ross returned to England, considerable difference of opinion existed as to the necessity for a farther search for Sir John Franklin; but Lady Franklin, who, through all her sad trials and sore disappointments, never wavered in her faith that he had accomplished the service he was sent to execute, nor hesitated at any sacrifice to effect the rescue of her husband and his companions, urged that the search should be resumed so strongly, and was backed by so many influential arctic authorities, that the Admiralty, to decide the question, ordered an arctic committee to sit upon the question of the resumption of the search for Franklin. That committee recommended that it should be again attempted, and, satisfied that Franklin was not at Melville Island, proposed that all the strength of such an expedition as might be sent should be employed up Wellington Channel, and, never dreaming of the Enterprise or Investigator having made such progress as they had done, merely advised as follows, touching support or aid to them :—

"With respect to the efforts now making to afford relief to the missing ships in the direction of Behring Strait, we do not venture to offer any suggestions, beyond a hope that, until further accounts are received from Captain Collinson and Commander M'Clure, the Plover may be kept fully provisioned" (in Behring Strait).

Now, according to the nature of the evidence before them—furnished mostly by the seniors in late expeditions, who took a very gloomy and unpromising view of the case—and considering also the amount of probability that any such navigation would be carried out as Captain M'Clure appeared to contemplate, in his admirable official letter from Behring Strait, there can be no doubt that this committee gave a correct opinion; but it was fortunate for England's naval history, and for humanity, that a parent's solicitude for a son absent in this expedition was more clear-sighted, and that by that, as well as by the zeal and energy of Lady Franklin, a modification of the official opinion was effected.

Mr Cresswell, indeed, in a letter which his kindness allows me to make public, pointed out, with almost prophetic vision, the very position in which the Investigator's crew was placed, and also suggested the means by which their wants might be supplied and their safety insured. The letter ran as follows:—

"Lynn, Norfolk, 23d March 1852.

"SIR,—In reference to the expedition now fitting out for the arctic regions under Sir Edward Belcher, I am induced earnestly to request the attention of the Lords Commissioners of the Admiralty to the following appeal:—

"In the sailing orders given to the expedition to Behring Strait, the sixteenth paragraph leaves to Captain Collinson the course to be pursued after leaving Point Barrow, referring him for assistance and direction to Captain Kellett, Sir W. E. Parry, and Captain Beechey.

"In the memoranda given by Sir Edward Parry and Captain Beechey, we find that they both consider the great object of the Behring Strait expedition to be to penetrate to Banks Land, or even to Melville Island.

"The writers of nearly all the letters relating to the position of Sir John Franklin consider the most likely quarter in which to find him to be the vicinity of Banks Land or Melville Island.

"Colonel Sabine, in a letter to Sir W. E. Parry, 15th June 1850, speaking of Sir John Franklin, says, 'His advance from Melville Island in the season of 1847 may have been limited to a distance of 50 or perhaps 100 miles at farthest,' and that 'in 1848 he may have endeavoured to retrace his steps, but only with partial success. It is, I apprehend, quite a conceivable case, that, under these circumstances, the crews, incapable of extricating the ships from the ice, may have at length been obliged to quit them and attempt a retreat, not towards the continent, because too distant, but to Melville Island, where certainly food (seals), and probably fuel, might be obtained, and *where they would naturally suppose that vessels despatched from England for their relief would in the first instance seek them.*'

"Captain Austin's expedition was directed, in accordance with the above authorities, to make its principal efforts in the vicinity of Melville Island.

"The intelligence of the fitting-out of Captain Austin's

expedition reached the Behring Strait expedition at the Sandwich Islands, and would unquestionably stimulate them in endeavouring at almost any risk to communicate with Captain Austin at Melville Island.

"It is desirable to realise as much as possible what would be the effect on the officers of the Behring Strait expedition on being possessed of the intelligence alluded to.

"There can be no reasonable doubt but that they would anticipate Captain Austin's expedition, aided as it was by steamers, reaching Melville Island, as Sir W. E. Parry had done so without steamers in 1819; and that Captain Austin would, as a matter of course, leave a supply of provisions and fuel, and possibly a boat, for any party that might reach that point from Behring Strait.

"*Trusting to this, a party might be pushed forward at imminent risk; but how terrible would be their despair on finding a barren notice of Lieutenant M'Clintock's visit in* 1851, *instead of the supply anticipated!*

"There is nothing unreasonable in the above supposition, as Captain Kellett in his evidence, page 170, says, '*Should Commander M'Clure be successful in getting far to the eastward, I am convinced, from a conversation I had with him, and indeed his own letter will show, that he will use every endeavour to reach Melville Island with his parties if he fail with his ship.*'

"Again, Captain Kellett, in a letter to Captain Collinson, 20th of May 1850, says: 'If you can pass Point Barrow and escape the shoal water, we shall see you come home by the Atlantic.

"Now the Investigator passed Point Barrow the 5th of August 1850, and nothing has been heard of her from that time; we may therefore conclude that she struggled to get to Melville Island through the seasons of 1850 and 1851. Certainly she did not return in 1851 through Behring Strait, or it must have been already known.

"If she is far to the eastward, in all probability Melville Island will be their only resource. Captain Collinson, in the Enterprise, passed Point Barrow in 1851, and will in all probability push to the eastward this summer, and may reach Melville Island.

"Under these circumstances, ought not the powerful arctic squadron now fitting out to have some reference to *the support of Captain Collinson's expedition*, as well as to following out Penny's discoveries, and to the faint hope of finding Sir John Franklin?

"I venture to suggest that one of the three steamers should proceed to Melville Island, if the navigation be open, as in 1819, leave provisions, fuel, and a boat at Winter Harbour: she could then return to the ships at Beechey Island.

"If the navigation should not be found open, then the steamer should make as far to the westward as possible, and send provisions to Byam Martin Island, or any favourable position that might be practicable, sending forward a party to Melville Island, to leave a record of their proceedings, to direct where to find provisions, and also to communicate the object of Sir Edward Belcher's expedition.

"If the Admiralty thought proper, in case of the steamer reaching Melville Island, she might winter there, and in the spring send searching-parties to the westward. They might be so arranged as to command a wide field of research, and possibly cross Captain Collinson or Commander M'Clure, as they would start from a point more than 200 miles west of Baillie Hamilton Island.

"I must apologise for trespassing upon you with such a long letter, but the importance of the object will, I trust, plead my excuse.—&c. &c.

"To Augustus Stafford, Esq., M.P.,
Secretary, Admiralty."

His Grace the Duke of Northumberland was then the Senior Lord of the Board of Admiralty. He was struck with the soundness of Mr Cresswell's views; and having sought the opinion of some arctic authorities upon the subject, a paragraph was inserted in the orders under which Captain Belcher sailed;* and his expedition, in consequence, became divided, for the twofold objects of seeking Franklin and affording aid to Captains M'Clure

* Paragraphs 5 and 6 of Captain Sir E. Belcher's instructions were as follows :—

5. "Arrived at this point (*Beechey Island*), two great objects will engage your attention :—

"First, the endeavouring to pass up Wellington Channel with one sailing vessel and one steamer; secondly, the advance of a similar force towards Melville Island.

6. "The object of the first of these expeditions will be, the endeavour to recover those traces of Sir John Franklin which ceased at Cape Bowden to the north of Beechey Island, and to follow up such traces, if they should be found. The object of the other expedition will be, to deposit, if possible, at Winter Harbour, Melville Island, or, failing that, at Byam Martin Island, a supply of provisions, fuel, and clothing, for any parties that might reach such positions from Captain Collinson's or Commander M'Clure's ships."

and Collinson. Two vessels were to go up Wellington Channel, under Captains Belcher and Osborn; whilst the Resolute and Intrepid, parting from them at Beechey Island, were, under the command of Captains Kellett and M'Clintock, to proceed to Melville Island; a duty which, so far as the last-named division was concerned, was, as will be seen, faithfully executed.

In accordance, therefore, with the recommendation of the Arctic Committee, and the instructions of the Lords Commissioners of the Admiralty, the expedition under Captains Sir E. Belcher and Henry Kellett, C.B., consisting of H.M.'s ships Resolute, Assistance, Pioneer, and Intrepid, left England in the spring of 1852. Without entering into a detail of the circumstances which brought about the eventual desertion of that fine squadron, it will suffice for our present purpose to say that on the 15th of August 1852, the Resolute, Captain Kellett, and the Intrepid, Commander M'Clintock, having on the previous day parted company from the Assistance and the Pioneer, proceeded towards Melville Island from the depot ship North Star, which was stationed at Beechey Island.

Melville Island was reached on the 1st September, and on the 5th instant the vessels made fast to the ice which yet filled up Winter Harbour, the well-known wintering-place of Sir Edward Parry in the year 1819. A depot of provisions was immediately formed; and unable to discover any cairn that had been erected by Captain M'Clure's party or that of Lieutenant Parks from the Enterprise, when they visited the place in the spring of the same year, the ships under Captain Kellett actually retired to a secure wintering-place under Dealy Island, some distance to the eastward, without being aware that Melville Island had been visited by those they were in search of, or that within 180 miles of them the Investigator was then lying in want of assistance!* Such are the difficulties and chances which prevent one party finding another in those regions.

Directly, however, that Captain Kellett was securely frozen in, he despatched autumnal parties of travellers to lay out provisions for the use of parties he purposed sending forth in the ensuing spring. One of these parties, under Lieutenant Mecham, left the ship on the 23d September, reached Winter Harbour, struck overland for Liddon Gulf, deposited provisions on its shore, and returned to Winter Harbour. Happening to inspect more closely than usual the famous mass of sandstone on which Parry had caused his ship's name to be engraved, Lieutenant Mecham could hardly credit his senses when he discovered a document upon its summit, detailing the accomplishment of the North-West Passage, and the position of H.M.S. Investigator in Banks Land! Great joy was there in Captain Kellett's squadron at having discovered a trace of one at least of the ships they sought; and many an anxious discussion took place at Melville Island whether they could next spring send off parties sufficiently early to reach the Bay of Mercy before Captain M'Clure might be obliged to abandon his

* As an additional proof of the difficulty of finding traces in the arctic regions, we may add that, although the low point of Melville Island, called Port Hearne, which was visited by Lieutenant Parks, has been repeatedly visited, and once expressly searched, his cairn or record has not been discovered.

ship: and, on the other hand, there were doubts whether he might not have been able to push on during the past summer, and perhaps have again to be sought in some new direction.

Impressed with the belief that the Investigator had got out of the Bay of Mercy, and passed to the north-west of Melville Island, Captain M'Clintock and Lieutenant Mecham, as seniors, chose routes which would intercept her supposed track; and consequently, the second lieutenant of the Resolute, Mr Bedford Pim, was, with Dr Domville, of the same ship, told off to start with sledges from Melville Island to Banks Land: and on March 10, 1853, they left their shipmates, amidst the prayers and cheers of all.

Meantime, April 1853 had come in on the Investigator. The retreating sledges were ready, the slender store of provisions was packed, those that were going strove to be sanguine, those that were to remain behind had written to cheer up mothers, wives, and sisters, who must have already begun to mourn their long absence. On the 5th of the month the first death since leaving England occurred on board the Investigator: it was occasioned by the thoughtlessness of the poor fellow himself, who, by way of a joke, went into the surgery and drank off the washings of several medicine bottles. But the moral effect of a death at such a time was distressing, and to re-inspirit the men their iron-nerved captain took an early opportunity of calling the crew together, and making an address to them in not ineloquent terms. In it he called their attention to the difficulties already mastered, to the honours won, to the grateful recompense their good service was certain to obtain for each, and to the merciful Providence which had so upheld them hitherto; and he begged them always to remember that, in the gloomiest hour of trial, relief might, and often did come, and that the darkest cloud had ever a silvery side to it. Cordially again did all assent to his opinions, and the poor fellows talked more cheerfully, and looked happier for what had been said to them.

The 6th of April 1853 came in. A fine deer was hung up ready to be cut up for the hearty meal that all hands were to partake of before their separation, which was to take place in the following week, when an event occurred which rescued them from further suffering and trials of fortitude. I give Captain M'Clure's journal almost verbatim upon this day.

"While walking near the ship, in conversation with the first lieutenant upon the subject of digging the grave for the man who died yesterday, and discussing how we could cut a grave in the ground whilst it was so hardly frozen—a subject naturally sad and depressing—we perceived a figure walking rapidly towards us from the rough ice at the entrance of the bay. From his pace and gestures we both naturally supposed at first that he was some one of our party pursued by a bear, but as we approached him doubts arose as to who it could be. He was certainly unlike any of our men; but recollecting that it was possible some one might be trying a new travelling-dress, preparatory to the departure of our sledges, and certain that no one else was near, we continued to advance. When within about two hundred yards of us, this strange figure threw up his arms, and made gesticulations resembling those used by Esquimaux, besides shouting, at the top of his voice, words which, from the wind and the intense excitement of the moment, sounded like a wild screech; and this brought us both fairly to a stand-still. The stranger came quietly on, and we saw that his face was as black as ebony, and really at the moment we might be pardoned for wondering whether he was a denizen of this or the other world; and had he but given us a glimpse of a tail or a cloven hoof, we should assuredly have taken to our legs: as it was, we gallantly stood our ground, and, had the skies fallen upon us, we could hardly have been more astonished than when the dark-faced stranger called out,—

"'I'm Lieutenant Pim, late of the Herald, and now in the Resolute. Captain Kellett is in her at Dealy Island!'

"To rush at and seize him by the hand was the first impulse, for the heart was too full for the tongue to speak. The announcement of relief being close at hand, when none was supposed to be even within the arctic circle, was too sudden, unexpected, and joyous for our minds to comprehend it at once. The news flew with lightning rapidity, the ship was all in commotion; the sick, forgetful of their maladies, leapt from their hammocks; the artificers dropped their tools, and the lower deck was cleared of men; for they all rushed for the hatchway to be assured that a stranger was actually amongst them, and that his tale was true. Despondency fled the ship, and Lieutenant Pim received a welcome—

pure, hearty, and grateful—that he will assuredly remember and cherish to the end of his days."

In a very short time the dog-sledge with two men arrived, and long and eager were the conversations and questionings which ensued. The Investigators felt perfectly bewildered with the rescue which had reached them just in time to save, in all probability, the lives of the thirty persons who were about to attempt to reach home with sledges and boats (as well as that forlorn hope intending to remain behind); and when the fact had perfectly realised itself to all, it may be imagined what their feelings were. It would be supererogatory to attempt to describe the fulness and gratitude of heart with which each must have thanked his God for all his mercies.

. . . . . .

CHAPTER XVII.

THE MIGRATION OF ANIMALS THEORY—IMPOSSIBILITY OF ARCTIC ANIMALS MIGRATING—HOW THE FACT BROKE UPON US—ACCUMULATIVE EVIDENCE—THE REINDEER OF THE ARCTIC ARCHIPELAGO—HERDING IN THE AUTUMN—CURIOSITY OF THE DOES—WONDERFUL PROVIDENCE OF NATURE—SLOW PROCESS OF DIGESTION—REINDEER HARASSED BY THE WOLVES—SCENT A SUBSTITUTE FOR SIGHT—THE WOLVES LESS HARASSING IN EARLY SPRING—MUSK-OXEN—THE HAUNTS OF THE OXEN—CAPTAIN MECHAM'S OBSERVATIONS—ADMIRABLE COATING OF THE MUSK-OX—MEASUREMENTS OF MUSK-OXEN—EXTRAORDINARY ACTIVITY OF THESE CREATURES—FEROCITY OF THE BULLS—GOAT-LIKE POWER OF CLIMBING—THE ARCTIC HARE AND ITS HABITS—THE ARCTIC LEMMING AND ITS HABITS—BEARS EAT THE LEMMING—THE PTARMIGAN.

IT will not here be out of place to throw together the observations generally collected upon the habits of those two interesting arctic animals, the reindeer and musk-ox. The facts are spread over a great amount of journalising, but the editor has carefully collated all new information and compared it with the observations of officers in other expeditions; and he alone is responsible for a theory involving the broad assertion that the animals of the vast lands north of America, within the arctic zone, do not migrate in the winter-time. For such a belief the editor and some others were nigh excommunicated as heretics in 1851, by some professional naturalists in the expedition then wintering under Griffith Island. But now that the trustworthy records of the voyages of Captains M'Clure, Austin, Kellett, Penny, Kane, and M'Clintock have put us in possession of data connected with the movements of the oceanic ice up to a very late period in the year, in different parts of the arctic archipelago, all are able to see that the theory of an autumnal migration of the herbivorous animals to the continent of America, for the purpose of avoiding the rigours of an arctic winter, is no longer tenable.

The great winter-drift, in 1849 and 1850, of Sir James Ross and Commander de Haven, from Barrow Strait and the top of Wellington Channel, proved that the ice around those lands was in motion long after the winter had set in, and that in the season of utter darkness, those wild, half-frozen seas were churning and rolling on in their mysterious course to southern latitudes. We have seen since then that the ice-beset Investigator in Prince of Wales Strait, and Captain Kellett's squadron in Melville Sound, were not stationary until the close of November; and long after that period, during spring-tides or in strong gales, there was abundant evidence that large spaces of water and weak ice existed around them; such, in short, as would be quite sufficient to prevent timorous deer or musk-oxen attempting a journey which would have puzzled even an amphibious animal. Additional testimony abounds elsewhere; the ice of Queen's Channel, and around the winter-quarters of H.M.S. Pioneer in Northumberland Sound (1852-53), was so weak and so heavily packed at the end of the winter, that it could with difficulty be traversed by our men; and near Dr Kane's winter-quarters, in Smith Sound, the ice was so treacherous that his parties could not cross it from Greenland to the western coast.

All these facts betoken insuperable difficulties in the way of an animal migration, simply from the absence of a highway for the poor brutes to pass from 78° to 68° north latitude, a distance of about 600 miles straight as the crow flies. On the other hand, we know that the reindeer winters in Greenland; for not even the most profound believer in an animal exodus has ever accused the poor creatures of embarking on the bosom of the waters of the Atlantic or Davis Strait, and swimming to Labrador; moreover, the Dutch and Russian fishermen, wintering in Spitzbergen in the old time, found reindeer there throughout the season; and lastly, from the abundant testimony of the officers of the Investigator during three winters, as well as those of our comrades at Melville Island in 1853, and our own observations during four winters in the arctic zone, we feel justified in saying that, beyond all doubt, the deer, musk-ox, hare, and lemming, do winter in the arctic archipelago.

It would not interest the reader to quote at length all the passages upon the subject from the different journals of officers lately engaged on arctic service; some remarked one fact, others another; yet by plodding over those ponderous arctic blue-books a very fair collection of data may be collected.

In the depth of the winter of 1850-51, deer, or recent traces of animals, were seen near the respective winter-quarters of Captain M'Clure, Captain Austin, and Captain Penny; and in the early spring of 1851, when the temperature was —40°, Lieutenant Aldrich observed reindeer, white as driven snow, grazing upon what he described as stony plains covered two feet deep with snow, and the animals so lean and winter-pulled, that no one could suppose they had been revelling on the American continent, and had just rushed up to 76° north to enjoy a low temperature and Lenten fare. They had their young fawns with them, which was an additional argument against a journey which, to and fro, could hardly be less than 2000 miles; and it is as well to remember that distance tells on animals as well as men.

Captains M'Clure and Kellett testify to these animals being found all the winter through about Banks Land and Melville Island. M'Clure's narrative contains several remarkable passages upon this head: we will give one, dated December 1852. "The deer have for the last few days," he says, "been coming from the southward to their winter quarters amongst the ravines and sandhills. Ninety have been met with at one time, and forty at another; but they are so wild that few have been shot. Our two seasons' experience shows that these animals do not migrate to the south, as is generally supposed, but bear the extreme rigour of the climate, and exist upon the scanty herbage here found, chiefly the dwarf willow, from off which they break the snow with their feet, and in doing so make a tapping noise that may be heard at a considerable distance when the weather is calm, frequently leading to their discovery by our sportsmen. The hares and ptarmigan have also descended from the high ground to the sea ridges, so that a fair supply is brought in."

In 1853, during a winter of bitter temperature, Captain G. H. Richards and ourselves landed on the north shore of Bathurst Land, and were not a little surprised to find that reindeer were very numerous on the uplands: they were browsing, with their fawns, upon a miserable vegetation which any other animal would have starved upon: the only plant which they did not appear to have touched was the saxifrage, notwithstanding that the young shoots or buds are remarkably sweet, and the favourite food of the ptarmigan.

That the reindeer crosses the firm ice of the archipelago in the spring, no one can deny; but it is in search of food, not to avoid a rigour of climate which nature has provided them with an admirable organisation to meet; but those tracks of deer, and sometimes the creatures themselves, have only been seen going in an easterly and westerly direction, between the islands of Melville, Bathurst, and Cornwallis, upon the one hand, and Melville, Eglintoun, and Prince Patrick, upon the other; but never in such numbers as to induce any one to call it a migration. No one has ever seen deer or any other herbivorous animal crossing Barrow Strait or Melville Strait, either going north or south. We will next touch upon the general habits of these wonderfully-constructed creatures, who, without any coating of blubber like the bear and the seal, are able to pass unscathed through a pitiless winter in a climate ranging, as far as yet known, from zero to —65°, a temperature which strikes like cold steel at the vital powers of a well-clad man, and rends iron and rock by its resistless power.

We will first speak of the reindeer. Their average

size and weight approximate to those of the ordinary fallow deer of our English parks. An exceptional case is sometimes seen in some lordly stag, who, though, like Tennyson's "many-wintered crow," admirably fitted to lead his herd, and forming a very fine object in an arctic solitude, would be uncommonly tough and strong eating anywhere but in 76° north latitude. They are by no means graceful creatures at any age; the joints are large and powerful in proportion to the size of the animal; the divided hoofs are very large, and from the animal being obliged to raise its feet high when going over the snow, its gallop has none of that beautiful spring which characterises the red deer of our isles, though the pace is a telling one, and soon carries the reindeer clear of anything but the long-winded long-legged wolf.

The stags cast their antlers, and the does drop their young, in May or June, about the time of the first thaws; the males and females are then not often found together, unless it be some gay Lothario, with half-a-dozen admiring spinsters—an exceptional case, however; and the female deer are at this season usually in small herds with their fawns; the little creatures—all eyes, ears, and legs—taking alarm at the slightest appearance of danger. The summer vegetation fattens the bucks and does amazingly, and the fawns grow apace, all three having a comparative holiday, and getting into condition to meet the trials of the coming winter; while the wolf and the fox, their sworn foes, are, during the same summer season, devoting their attention to the infant seals and bears, or attending to their own little domestic duties. Indeed, in the height of the arctic summer, the swampy state of the lowlands and the cutting effect of the stony hills, as shown in the state of our poor dogs' feet and our own boots, was strong testimony against the wolf or fox being able to do much at that season against hoofed animals. As the autumn frost sets in, and the snow again spreads its pall over the death-like scenery of the north, the wolves again return to watch and harass the unfortunate deer.

For mutual protection and warmth, and following the natural instincts of gregarious animals, the deer in October commence to troop together, forming large herds of bucks, does, and fawns. Some have been counted numbering 60 head. The stags are evidently responsible for the discipline of these large herds, as well as their safety: upon the latter head, Captain Mecham tells an amusing anecdote.

In October 1852, he was crossing that portion of Melville Island which intervenes between Liddon Gulf and Winter Harbour, and fell in with as many as 300 head of deer: indeed, he says reindeer were always in sight in herds varying from 10 to 60 in number. One of these herds, containing 20 head, he tried to stalk up to on the 7th October, but failed in getting a shot at them; for although the does, with the inherent failing of their sex, were extremely curious, and made one or two efforts to escape from the herds, and examine the "strange gentleman," the stags would in nowise tolerate such conduct, chastised them rather soundly with their antlers, and kept the herd together and moving, by running rapidly round and round, uttering at the same time a strange noise, which seemed to alarm the herd, and keep it flying from the suspected danger.

The coats of these creatures, which during summer becomes remarkably thin, and adapted admirably in colour to that of the snow-denuded soil, now rapidly thickens and again returns to its pristine whiteness. It is not a fur, in the strict acceptance of the term, but it forms an admirable non-conducting substance.

As winter advances, and food requires to be sought over large areas, the herds break into parties of 10 to 20 animals. The various lichens, a species of tripe-de-roche, the sprouts of the ground willow, as well as Iceland moss, are their principal food; but it must be remembered that arctic vegetation has no time in the autumn to wither or decay; whilst in full bloom, and before the juices have time to return into the parent root or be otherwise dissipated, the magic hand of the Frost King strikes them, and thus the wisdom of the Creator has provided for the nourishment of His creatures fresh and warmth-creating food, lying hid under a mantle of snow, which the instinct of arctic animals teaches them to remove and so feed upon the stores beneficently preserved beneath.

There is another peculiarity worthy of notice. Most herbivorous animals have a slow system of digestion, even in a domestic state; our cattle and sheep, for instance. This is still more the case in the musk-oxen, reindeer, and arctic hare, and is of infinite use in lands where the vegetation is scanty and widespread, and the weather occasionally so severe as to oblige these creatures for two or three days at a time to look purely to their own safety by seeking shelter from the snow-storms in deep ravines or under lofty cliffs. It appears in their case as if nature extracted from their food a greater quantity of nourishment than she does from that of animals in more southern latitudes, or it may be that the retention in the stomach or intestines of vegetable matter after all nutriment has been extracted, serves to check the cravings of their appetite.

The majority of the musk-oxen and deer, the former especially, had their entrails distended with food (apparently quite digested), whilst the country around, in many cases, was as barren as a macadamised road, fairly leading to the inference that these creatures must have been a long time collecting what they had within them, and that it had been a long time swallowed. It struck us likewise that it required the vital principle of the animals to be in full activity to prevent such food from becoming a source of disease; for if a musk-ox was shot, and left a short time without being disembowelled, the flesh became tainted throughout with a strong odour, rendering it uneatable.

Another strong fact which bears upon the impunity with which these creatures can winter in high latitudes, is, that in Lapland, where these reindeer are used for tractile purposes, it is considered quite enough food for a working animal if they are able to give it daily four pounds of lichen (*Cenomyce rangiferina*); and on that dietary a reindeer will be in sufficiently good condition to go occasionally without food for two or three days, and does it without apparent distress.

So far as food is concerned, and an organisation fitted to meet the extreme temperature of the pole, reindeer are thus amply provided; but their sorest trial must be the constant rapacity of the wolves which are ever hanging round them throughout the winter season. As the season advances, the reindeer appears to resign itself to this inevitable social misery; and the cool manner in which a small flock of them may be seen grazing with an *entourage* of half-a-dozen hungry wolves is very strange, and evinces, to say the least of it, great philosophy.

A herd of deer thus surrounded by the wolves was often seen by arctic sportsmen. The wolves, far too great cowards to rush in upon their prey, would endeavour to startle the herd by their long-drawn unearthly howls: sometimes a frightened deer, horror-stricken at the abominable chant, would dash madly from the herd, and thus leave the protection of the guardian stags,—away all or a portion of the wolfish fraternity go after it. In many cases, the scene might be briefly summed up with the old three-volume *dénouement* of—a rush—a shriek—a craunching of bones, and snarling of beasts of prey—and all is over! for the wonderful powers of swallow and horrid voracity of an arctic wolf must be seen to be understood; no writer would peril his reputation for veracity by repeating what has been seen on that head. Sometimes, however, the frightened deer would gain the open country, and goes a wonderful distance dogged by the persevering wolf, who assuredly kills it, unless another herd is met which admits the hunted deer into its ranks.

Occasionally, a herd of deer are grazing, and one of them hits upon a spot where the food is plentiful; it naturally lingers there, whilst the herd moves slowly on against the wind, as is their habit. The wolves immediately mark the straggler, and stealthily crawl up, their object being to cut him off from the herd: that effected, there is a howl and rush, and its fate is instantly sealed if the deer does not evade its foes by extraordinary exertions.

All through the winter these scenes go on, scent serving the creatures when sight is useless; and many a sportsman, in the December darkness of the Bay of God's Mercy, often wished his olfactory nerves were as sensitive as those of the wolf, for, although he could hear the deer, it was then impossible to see them; and many a bad shot was made by a hungry man at a large pair of sorrowful eyes which loomed out of the mist around, for the sportsmen could not tell whether the deer was two or twenty yards from him.

During the depth of the severe winter of 1852-53, the deer approached close to the Investigator: and in doing so they quitted the land. Whether this was done with a view of seeking the warmth which instinct told them was being given out by the ship, which was a perfect volcano compared to the bitter temperature everywhere prevailing—95° below the freezing-point of water—or whether it was for security against the wolves, it is difficult to say, but most probably from the former cause; for we remember that the foxes of Leopold Harbour, in 1848, soon discovered the warmth thrown out by the squadron under Sir James Ross, and wisely burrowed and bred in the snow embankments thrown up around the ships.

Winter, with its sore trials, has of course its limits;

and it is astonishing how early in the new year relief comes to the harassed reindeer. In February and March the seals commence breeding, and their helpless young are luscious morsels, calculated to attract the attention of all beasts of prey. The reindeer's holidays then commence, as well as those of the arctic hare and lemming, who contribute their quota to the winter meals of wolf, fox, and bear.

As spring advances, the herds gradually disperse, and the deer may then be seen in twos or threes, until, as I have before said, the autumn again approaches.

The general habits of the musk-oxen of the archipelago resemble strongly those of the reindeer; but they appear to be principally confined to Melville Island and Banks Land.

None of them were seen alive on Bathurst or Cornwallis Land, although ancient skulls and bones have been found on both shores of Wellington and the Queen's Channel. One musk-ox was found in 1851, in Byam Martin Island; it appeared to have died from old age or starvation. Captain M'Clure only obtained three oxen from Mercy Bay, but subsequent visitors to Banks Land, Captain Mecham and Mr Krabbé, saw numbers. So far as places visited can be taken as an authority upon the subject of their locality, it appears as if the south-west extreme of Melville Island was their favourite haunt, especially that portion of it lying between Liddon Gulf and Cape Russell; and it is worthy of remark, that that portion of Melville Island, although possessing a southern aspect, impinges upon the vast area of never-thawing ice, that "land of the white bear," from whence the west wind appeared to bring such intense cold whilst the Investigator was imprisoned against Ballast Beach in Banks Land.

Commander G. F. Mecham, whose interesting remarks, whilst searching in the above direction, are of great value in many respects, makes the following general observations upon the animals he saw in 1853 :—

"Game was only shot when required for our use, otherwise great quantities might have been obtained on Melville Island. About the sloping land from Cape Smyth up to the head of Liddon Gulf, animals were seen in great numbers, but particularly about the 115° of longitude, where, both in April and June, musk-oxen were very numerous. I saw, in a walk overland of ten miles, as many as 150 head of cattle. At Cape Smyth, on June 18th, a perfectly white musk-cow was seen with a black calf grazing with another cow and calf of the usual colour. Only one small herd of reindeer was seen while crossing Melville Island to Winter Harbour in July, as the land was then covered with water, or else in a deep swamp. In June and July, innumerable lemmings were seen both on the land and ice. Those on the ice were frequently carried off by the burgomasters, which were always in great numbers wherever the land was high or steep. At the entrance of Liddon Gulf two large flocks of snowy geese were seen, but, in general, all the water-fowl were very wild.

"From the barren state of the soil of Eglintoun and Prince Patrick Land, I am inclined to think that it is not a very favourite resort for animals. Several traces were seen in May and June on the ice, all travelling *from* Melville Island to the westward. On Patrick Land we found vegetation only immediately on the south beach, and that only as far as 122° W. Throughout the journey beyond that, until returning again to the southern shore on June 1st, no traces or animals of any kind were seen, except two bears off Cape Manning.

"The musk-oxen were all very wild in April, and generally seen in large herds from ten to seventy in number. In June they were stupidly tame, and seemed to be worried with their heavy coats of wool, which were hanging loosely down their shoulders and rumps in large quantities; the herds much smaller, and generally composed of cows and calves.

"At Cape Russell I walked up to within ten yards of two cows and a bull without their taking the least notice of me, and when I fired, only ran about five yards and commenced grazing. The cows were at first butting at the bull, who received their blows with the crown of his horns, which sounded like the meeting of two heavy skittle-balls."

The heavy coat of wool with which the musk-oxen are provided, is a perfect protection against any temperature. It consists of a long fine black hair, and in some cases white (for it is not ascertained that these oxen change their colour during the winter), with a beautiful fine wool or fur underneath, softer and richer than the finest alpaca wool, as well as much longer in the staple. This mantle appears to touch the ground, and the little creatures look like a bale of black wool, mounted on four short nervous goat-like legs, with two very bright eyes, and a pair of sharp wicked-shaped horns peering out of one end of it. Captain M'Clintock, of H.M.S. Intrepid, gives the following dimensions of some oxen shot by him in 1851, which are a very fair average, the animals seldom exceeding the size of Shetland ponies. [See Table, p. 237.]

They seem to be of very uncertain temper, sometimes standing stupidly glaring at their assailants, whetting their horns against their forelegs; and at other times our sportsmen had to be quick in escaping from their fury.

Of their activity when excited Captain Mecham speaks in another part of his diary, before quoted.

"During our stay, I proceeded to the northward, overland towards the head of Hardy Bay, Melville Island. The land rises to an elevation of about 800 feet above the sea, and nearly all the hills are of a remarkable table-shape. Musk-oxen are here in very great numbers; on one plain I observed as many as seventy grazing within a circuit of two miles. On my approach they divided into herds of about fifteen each, headed by two or three enormous bulls. Their manœuvres were so quick and regular that they were more to be compared to squadrons of cavalry than anything I could think of. One herd advanced several times at a gallop within rifle-shot, and formed in perfect line with bulls in advance, showing a formidable front of horns. The last time they advanced at a gallop to about sixty yards, and formed in line, the bulls at the same time snorting and tearing up the snow. Immediately I fired they wheeled round, joined the main herd, and made off out of sight, only waiting occasionally for the wounded one."

	Measurements of Musk-Oxen.	Bull.	Bull.	Cow.	Cow.	
Shot at Melville Island in 1851.		ft. in.	ft. in.	ft. in.	ft. in.	The roots of the horns meet over the forehead, and in the bulls spread out longitudinally, forming a secure shield for the head. In the cows the roots or bases of the horns are much smaller, and are buried in long hair.
	From base of horns to root of tail .	7 2	5 10½	5 2	5 4	
	Width of the base of horns measured longitudinally	...	0 11½	...	...	
	Base of horns (including their width) to the nose . . .	1 10½	...	1 4½	...	
	Hoof to tip of shoulder . . .	4 9	4 7	4 1½	...	
	Hind hoof to top of rump . .	4 3	...	...	...	
	Corner of mouth to eye . .	0 10	0 10	0 9	...	
	Round of muzzle above the nostrils	1 9	1 9	1 7	...	
	From one eye to the other . .	0 11¾	...	0 9	...	
	Round of fore leg, just above the hoof (fetlock ?)	0 7	0 7½	...	...	
	Width of fore hoof . . .	0 4½	0 4½	0 3¾	...	
	Its circumference	1 4	1 3¼	1 1	...	
	Width of hind hoof . . .	0 3¾	...	0 3¼	...	
	From tip to tip of horns . .	2 8	2 3¼	1 3½	...	
	Length of each horn . . .	2 3	2 0	1 7	...	
	Length of tail	0 2	...	0 1¾	...	

And in Captain L. M'Clintock's sledge-journey along the northern coast of Melville Island and Prince Patrick Island, he gives a glowing description of an encounter with a noble bull, which we transcribe as it stands in the Blue-Book of 1855 :—

"We saw and shot two very large musk-bulls, a well-timed supply, as the last of the venison was used this morning; we found them to be in better condition than any we had ever seen. I shall never forget the death-struggle of one of these noble bulls; a Spanish bull-fight gives no idea of it, and even the slaughter of the bear is tame in comparison. This animal was shot through the lungs, and blood gushed from his nostrils upon the snow. As it stood fiercely watching us, prepared yet unable to charge, its small but fixed glaring eyes were almost concealed by masses of shaggy hair, and its whole frame was fearfully convulsed with agony; the tremulous motion was communicated to its enormous covering of tangled wool and hair; even the coarse thick mane seemed to rise indignant, and slowly waved from side to side. It seemed as if the very fury of its passion was pent up within it for one final—a revengeful—charge. There was no roaring—the majestic beast was dumb; but the wild gleam of savage fire which shot from his eyes, and his menacing attitude, was far more terrible than the most hideous bellow. We watched in silence, for time was doing our work, nor did we venture to lower our guns until, his strength becoming exhausted, he reeled and fell.

"I have never witnessed such an intensity of rage, nor imagined for one moment that such an apparently stupid brute, under any circumstances of pain and passion, could have presented such a truly appalling spectacle. It is almost impossible to conceive a more terrific sight than that which was presented to us in the dying moments of this matchless denizen of these northern wilds."

It appears to be doubtful whether the wolf, naturally a most cowardly creature, is able to act in any way offensively against the musk-oxen; the general impression amongst the naval officers employed in localities where a good opinion upon the subject could be formed, was, that the wolf could only attack the lame or sickly cattle.

The activity of these oxen, and goat-like power of climbing, is very remarkable, and much at variance with their clumsy appearance. They have been seen making their way, when frightened, up the face of a cliff which defied all human efforts to follow them, and going down the precipitous sides of ravines by alternately sliding upon their hams, or pitching and arresting their downward course by the use of the magnificent shield of horn which spreads across their foreheads, in a manner to call forth the astonishment of the beholder.

The arctic hare (*Lepus glacialis*) collects in herds or troops during the fall of the year, in the same manner as the deer. Two hundred of them have been met at a time; and at one of their favourite haunts, Cape Dundas, Melville Island, a complete highway, three yards broad, was seen, the tread of their numbers having beaten the snow perfectly hard. In winter they burrow under the snow for protection, as well as to seek their food. Captain M'Clintock says, "They are everywhere found, but of course most numerous where the pasture is most abundant, as on Banks Land and Melville Island." The sportsmen of the Resolute and Intrepid shot 161 hares in a twelvemonth on the latter island; their average weight, when fit for the table, was 7 lb., and from 10 lb. to 12 lb. with skin and offal. During summer the hare, as well as the lemming, seeks protection from beasts of prey under large boulders of rock, or in the face of rocky ravines. The hares in summer have been seen in groups of from twelve to twenty in number. Their skin is so delicate, that although the winter fur is very beautiful, and the colour a brilliant white, it cannot be applied to any useful purpose. They do not hybernate, and, strange to say, the Investigators generally found them amongst the heavy hummocks of the floe-ice in Mercy Bay, as if flying to that rugged ground from the wolves or foxes. They differ from the European hare in bringing forth five or seven young at a birth.

That interesting little creature, of the order Rodentia, the arctic lemming (*Myodus lemmus*)—a perfect diamond edition of the guinea-pig—is very like the hare in its habits, but is generally found in large families. They have been seen at all seasons, and in winter are perfectly white; but feeding and living as they do under the snow, it is only the keen-nosed fox or Esquimaux dog that can detect their position and enjoy the sweet morsel they afford.

In summer, about the end of May, or early in June, they have a peculiar habit of going off the land on to the frozen surface of the sea. They do not seem to have any definite object in doing so, and cannot be said to be migrating. Possibly the thaws induce them to leave the land: the seamen, in their quaint way, used to say, "Them blessed little lemmings must be arter salt, I should think, sir!" and really there seemed to be no other way of accounting for their presence on the floe at such a season of the year. We found them steering off shore from the north coast of Melville Island, leaving comparative plenty behind them, and as far as the eye could see on a clear day from land of considerable height there was nothing in the shape of *terra firma* in the direction whither they were bound. When thus exposed upon the open floe, owls, gulls, and foxes pick them up for food. Can it be that Providence occasions this exodus for the purpose of feeding these creatures, and thinning down the numbers of an animal which would otherwise multiply exceedingly, and eat up all the vegetation of a naturally sterile region?

One would hardly suppose so tiny a creature would serve as food to such large animals as polar bears, but that it is so, the following extract from my journal will show, the place referred to being a valley on the north-east side of Bathurst Land debouching into Queen's Channel:—

"*Saturday, July* 2, A.M.—Saw some shoals and the *Cub* and *Bear* islets to seaward. Made sail to a rattling breeze, and, favoured by the ice, we went along at a good pace until 3.30 A.M., when, seeing some drift-wood lying about, which it was important should be examined, I halted and encamped, dispersing the men along the beach to bring all in they could find. Walking landward to obtain a view from a hill, I was startled to see a she-bear and two cubs some distance inland. Watching them carefully, I was not a little interested to see the mother applying her gigantic muscular power to turning over large blocks of sandstone, under which the unlucky lemmings at this season take shelter. Directly the she-bear lifted the stones, which she did by sitting upon her hams and pulling them towards her with her fore-paws, the cubs rushed in and seized their prey, tossing them up in the air in their wantonness. After repeating this operation until her young must have made a very good meal, I was glad to witness the bear's mode of suckling her young—a sight, I should think, rarely seen. Seated on her haunches, with the backbone arched, so as to bring the breasts (which were situated between the shoulders) as low as possible, the youngsters sucked away in a standing attitude. Anxious to secure this family party, we proceeded to burn all sorts of strong-smelling delicacies; and at last she brought her babes down, though very warily, and when more than 100 yards off, turned away, evidently suspicious. Following her I contrived, at about 150 yards, to pass a ball (Minié) through her body, abaft the shoulder. The cubs at once made off, though I should think they had not long been born, being about the size of an Irish retriever. Joined by a couple of the men (Hall and Wicketts), who soon outstripped me, we eventually, after a long chase, came up with her; the brute, seeing she could not escape, had apparently made up her mind to wait for us behind a range of hummocks. When close to her, I learnt that they had one shot each left in their guns; but as the men longed to go in at her, we walked up, the brute most artfully hiding her body so as to get us within reach for her rush. The wonderful similarity of colour between the fur of the bear and the snow facilitated her manœuvre, and we were within thirty yards of her when she rose. It was a ticklish moment, for the brute was venomous from desperation. The men behaved very coolly, however, merely saying to one another, "Steady!" Hall fired, but only grazed her; she still came on, when George Wicketts, with my Minié (which I thought he was fully entitled to fire after so successfully bringing the brute to bay), struck her smartly in the fore-shoulder. With a snap of the teeth, which it was satisfactory to know was not on ourselves, she turned round, and staggering along, fell dead into her lair again; and we returned to the boat to send the small sledge for the blubber.

"The she-bear was miserably lean, nothing in her stomach, and her skin in poor condition. Whilst they were skinning her, the poor little whelps ran up to be suckled; the men tried to catch them; failing in that, knocked their brains out; their little stomachs were perfectly distended with the unfortunate lemmings, which they had swallowed entire."

Perhaps the most curious fact of all connected with the existence of animal life in high latitudes during the most severe temperatures is, that ptarmigan were found throughout the winter in Melville Island and Banks Land. I might also add, that they have likewise been seen by officers who wintered at Beechey Island, where a small covey was flushed in the depth of the winter 1852-53. But it is best on this point to give the remarks of an officer who has had admirable opportunities of observing the fact—Captain F. L. M'Clintock, R. N.; he says:—

"The willow-grouse never goes north of Banks Land, the common ptarmigan (*Tetrao lagopus*) is the only bird of that species found on Melville Island. They are most numerous in April, generally in pairs, and in September they collect into coveys, sometimes of as many as fifteen or twenty birds, previous to a flight southward. After that month a few were seen, and those were birds which probably had not paired during the previous season. Some ptarmigan were shot in January and February, in excellent condition; of these the largest weighed 2½ lb., and its crop contained 2½ ounces of the slender shoots of the dwarf willow; many of these shoots were as thick as a crowquill, and ¾ inch long: when ready for cooking the bird weighed 1¼ lb.; no starveling this! Six hundred and eighty-four ptarmigan were shot on Melville Island in twelve months, by the people of the Resolute and Intrepid, being more than the Investigators got altogether." There is reason to believe that these hardy birds burrow under the snow for warmth, protection, and food, as the hares and lemmings do.

The snowy owl, as well as the raven, brave the winter and darkness of the arctic zone. We have related how the crew of the Investigator used to watch for their friend Ralpho. Another expedition, beset in Wellington Channel during the winter of 1853-54, likewise observed the black form of the arctic raven wheeling slowly over their ships; it often came in the wake of a gaunt white wolf which haunted us, and was known as the "ghost," from the impossibility of getting a shot at it. No doubt, the presence of a ship full of men, and the quantity of scraps and offal, might have a good deal to do with such a bird remaining in our vicinity; but some of our sledge-parties which were despatched directly the light came back, but still during the most rigorous season of the year, observed ravens in the neighbourhood of the high precipices or bluffs common to those regions. We concluded, therefore, that they frequented such places as positions of safety from the arctic fox, and that on the higher ridges they picked up sufficient food, in the shape of ptarmigan and lemmings, to satisfy their hunger.

The late Admiral Sir John Ross, during his long and trying sojourn in Boothia, observed the ravens digging through the snow to reach some tit-bit which their keen scent could have alone told them was hid away beneath. These birds, it is however as well to tell the reader, are by no means common in the desert regions of which we are writing; perhaps a dozen of them only were seen by our navigators in Barrow Strait and Wellington Channel during four or five years we spent there.

We must now take leave of this interesting subject, and return to the Investigator in Mercy Bay, Banks Land.

CHAPTER XVIII.

CAPTAIN M'CLURE PROCEEDS TO MELVILLE ISLAND TO SEE CAPTAIN KELLETT — M'CLURE'S VIEWS — HIS LETTER — CAPTAIN KELLETT ONLY GIVES LEAVE FOR HEALTHY VOLUNTEERS TO REMAIN IN INVESTIGATOR—MEDICAL SURVEY UNSATISFACTORY —ABANDONMENT OF H.M.S. INVESTIGATOR—DEPOT OF PROVISIONS FORMED AT MERCY BAY — ARRIVAL ON BOARD H.M.S. RESOLUTE AND INTREPID — SEARCHING PARTIES UNDER CAPTAIN M'CLINTOCK, LIEUTENANTS MECHAM AND HAMILTON, UNSUCCESSFUL — CAPTAIN KELLETT DECIDES UPON FALLING BACK UPON BEECHEY ISLAND — UNFAVOURABLE SEASON — SQUADRON BLOWN OUT OF WINTER-QUARTERS—ARRESTED AT BYAM MARTIN CHANNEL — LARGE SUPPLY OF FRESH MEAT PROCURED — RESOLUTE AND INTREPID CAUGHT IN THE PACK —WINTER IN THE PACK—THE PHŒNIX ARRIVES AT BEECHEY ISLAND, AND TAKES HOME LIEUTENANT CRESSWELL—1853— THE INVESTIGATORS PASS A FOURTH WINTER WITH IMPUNITY, AND THEN LEAVE THE RESOLUTE FOR BEECHEY ISLAND—THE LAST OF THE INVESTIGATOR—CAPTAIN SIR EDWARD BELCHER ORDERS THE RESOLUTE AND INTREPID TO BE ABANDONED — NEWS OF COLLINSON, IN 1852, HAVING PUSHED ON INTO THE ICE—ASSISTANCE AND PIONEER ORDERED TO BE DESERTED— PHŒNIX AND TALBOT ARRIVE WITH PROVISIONS AND FRESH CREWS—ALL RETURN HOME—INVESTIGATORS REWARDED IN 1855.

Directly the first joy and excitement arising from the pleasing news of another expedition being at hand to assist them in reaching England by way of the Atlantic had passed off, it became necessary to act quickly before the season of sledge-travelling had passed away in Melville Strait. Captain M'Clure decided upon going at once to see Captain Kellett, and make arrangements with him for having all his sickly hands sent to Melville Island, and thence home. He still adhered to his original plan, which was to carry home the Investigator by waiting in her through another arctic summer and winter (that of 1853-54), before abandoning her, and retreating to Melville Island. With this plan in view, he penned the following letter; and remarkable as the whole tone of it is, I should do injustice to the cool, unflinching nerve of the writer, did I not place in italics that paragraph in which, with generous heroism, he points out the inutility of risking more lives should he and his ship again be missing.

To the Secretary of the Admiralty.

"Her Majesty's Discovery Ship-Investigator,
Bay of Mercy, Baring Island (now Banks Land),
April 10, 1853.

"SIR,—In the event of our not getting to England this year, I think it necessary to acquaint you, for the information of the Lords Commissioners of the Admiralty, what our operations will be in 1854, that their Lordships may be enabled to take such co-operative measures for our relief as may appear expedient.

"Should the ice break up in this bay sufficiently early to permit our getting through the straits this season, and finding the water open to the eastward of Leopold Island, it would be my object to push forward, without stopping to take on board any provisions from Port Leopold; but if, on the contrary, the ice should be thick towards Lancaster Sound, I would, if possible, proceed to Port Leopold, and complete a twelvemonth's provisions, and then risk wintering in the pack, or getting through, in preference to remaining at the above port; if, however, we are detained in this bay till next year, it will then be requisite to leave towards the end of April and make for Port Leopold, where I am aware that there is a good boat, a house, and ample supplies; and, when the navigable season opens, proceed to Ponds Bay, coasting along the shore of Barrow Strait; arriving at Ponds Bay, and if finding from the Esquimaux that no whalers have as yet been there, I should there await their appearance as long as my provisions would admit, and then go down the west shore of Baffin Bay, keeping close along the land floe, where whalers or their boats are almost certain of being met with; failing this, I should cross to Disco, with the hope of getting a passage in one of the Danish vessels which come there annually, and leave about the beginning of September; or being too late for them, either charter or purchase one of their coasting schooners, which I believe trade amongst the settlements, if she was capable of standing an Atlantic voyage; could neither of these be accomplished, we must of necessity remain until the following season at that settlement.

"Should any of her Majesty's ships be sent for our relief, and we have quitted Port Leopold, a notice containing information of our route will be left at the door of the house on Whaler Point, or on some conspicuous position; *if, however, on the contrary, no intimation should be found of our having been there, it may at once be surmised that some fatal catastrophe has happened, either from our being carried into the Polar Sea or smashed in Barrow Strait, and no survivors left. If such be the case, which, however, I will not anticipate, it will then be quite unnecessary to penetrate farther to the westward for our relief, as by the period that any vessel could reach that port we must, from want of provisions, all have perished; in such a case, I would submit that the officer may be directed to return, and by no means incur the danger of losing other lives in quest of those who will then be no more.*

"As, however, it may occur (as was the case with Sir John Ross) that the ice may not break up in Regent Inlet during the whole summer, it is as well to provide against such a contingency; if such should happen, it would be necessary to winter at Port Leopold, unless apprised of the locality of any ship that might be sent for our relief, which I think might be accomplished without any great difficulty, as although such vessel may not be enabled to get far up the straits, yet as Admiralty Inlet would be pretty certain of being clear of ice, she might proceed thither, and in some secure bay freeze in, and when the straits were firmly frozen, about the middle of October, a small travelling party could be despatched with the intelligence; the whole would then proceed to her, and although rather late in the season, men working for their lives are not likely to be discouraged by a little cold.

"Whatever may be the final termination of this long, tedious, but I hope not unimportant voyage, I beg, Sir, that you will assure their Lordships that in every stage

I have been guided entirely by what I have considered to be my duty in prosecuting to the utmost the object for which the expedition was fitted out; and although we have not succeeded in obtaining any information which could afford the slightest clue to the fate of our missing countrymen, I hope that the services performed in the tracing of a very great extent of coast-line, the discovery of much new land (a portion of which is inhabited by a simple and primitive people not hitherto known), and, above all, the accurate knowledge of that 'Passage between the Atlantic and Pacific Oceans,' which for so many hundred years has baffled maritime Europe, its very existence being almost considered doubtful, will, I trust, be considered events sufficiently interesting and important to elicit from their Lordships a favourable consideration of our services.—I have the honour to be, Sir, your obedient and humble servant,

(Signed) "ROBERT M'CLURE,
Commander."

In accordance with his determination, Captain M'Clure left the Investigator with a sledge-party, and reached, the 19th April, the Resolute and Intrepid, commanded by Captain H. Kellett, C.B., and Commander Leopold M'Clintock. We can appreciate the meeting between himself and Captain Kellett, for they had parted on that eventful day in Behring Strait in 1850, when the latter gave Captain M'Clure an opportunity of waiting for his consort, which had he accepted, the North-West Passage would not have been achieved, and the search for Franklin would not have been completed upon even one line from ocean to ocean. Captain Kellett at first concurred in M'Clure's desire to save the Investigator; but when Lieutenant Cresswell of that ship arrived on 2d May, with some of the most weakly hands, and reported that two more deaths had already occurred in the Bay of Mercy, Captain Kellett, as senior officer, felt that his responsibility would be great if he allowed the zeal of Captain M'Clure or his followers, in fulfilling the requirements of professional honour, to jeopardise the lives of those who had so gallantly done their duty.* It was

* The state of the men brought over by Lieutenant Cresswell is best described in the following evidence given by Dr Domville before the Select Committee of the House of Commons, which assembled, in July 1855, to decide upon the reward due to Captain M'Clure and his men.

Captain M'Clure arrived on the 19th of April at her Majesty's ship Resolute, and he remained on board until the 2d of May, when another party from his ship arrived. "Until this period Commander M'Clure had been detained by Captain Kellett, the defective condition of his sledge's crew (who had doubtless been selected as the most efficient) being such as to cause some apprehension for the capabilities of the remainder to make a further sojourn in these regions; and most forcibly did the appearance of the above detachment justify the measure. Some vague information of their enfeebled condition had preceded them; the stern reality now presented itself: one officer subject to periods of mental aberration; one man in a state of *dementia* (or imbecility), his condition and appearance rendered still more pitiable from severe frostbite of the fingers; two men carried on the sledges, the one with scurvy, the other with urinary disease and phlegmonous inflammation of the leg; the remainder all more or less affected with scorbutic disease and debility, as indicated to the spectator in the tottering gait, attenuated form, and careworn expression of countenance, occasionally lighted up as the truth and recollection of their altered condition flitted across the imagination; a change (as some expressed themselves) difficult to realise. For several months past their thoughts had been pregnant with the uncertainty of the future, to which no definite results could be assigned."

then arranged that Dr Domville, of the Resolute, should return with Captain M'Clure to the Investigator, hold a medical survey on every person on board of her, order those home who might not be considered fit to withstand another winter, and then give the healthy the option of remaining in her for a fourth season, or not, as they might choose. On reaching the Investigator, Captain M'Clure addressed his men relative to their volunteering to remain out, and then gave twenty-four hours for the medical survey to take place, and for the sound men to make up their minds whether they would stay with him or go home. The survey, however, was fatal to the hopes of the resolute leader; for on the following day, May 23, he writes in his diary, "My surprise and mortification at finding only four men who felt able to go through another winter were great, but I must do all my officers the credit to say that they came most nobly and spiritedly forward, tendering their services, and expressing anxiety to remain and abide the chances of another season." Admiring as we may the iron will of M'Clure, still on the whole it was best that circumstances did not allow him to remain; for the medical report was most serious, and, as it subsequently turned out that the season of 1853 was an unfavourable one at Melville Island, and the Investigator could not have escaped, we may rejoice that these gallant men were put to no further trials in Mercy Bay.

The report of the medical men was to the following effect:—

"REPORT OF SURVEY OF CREW OF INVESTIGATOR.

"Her Majesty's Ship Investigator,
Bay of Mercy, 23d May 1853.

"SIR,—In obedience to your orders conveyed to us through Commander M'Clure, directing a survey to be held on the officers and men remaining on board this ship, with a view of ascertaining their general state of health and efficiency for further service in the Polar Sea, we have the honour to inform you that we have this day held a strict and careful survey accordingly, and beg to state, as the result thereof, that their present state of health is such as renders them utterly unfit to undergo the rigour of another winter in this climate, without entertaining the most serious apprehension for the consequence.

"There exists in all of them at present, with one or two exceptions, well-marked evidence of scurvy and debility in various stages of development, with great loss of flesh and strength, as may be seen from the remarks appended to the name of each in the accompanying list, which calls for their departure from these regions as early as possible as a matter of urgent necessity, and the adoption of prompt means to insure the same, that they may be placed under the salutary influence of such antiscorbutic and other agents as are essential for their recovery and ultimate safety.

"We are also of opinion that the reduced allowance of provisions on which they have been victualled for a period of nearly twenty months is one which we consider, and the past experience of others has likewise proved, to be quite inadequate for maintaining the health of the men exposed to the rigorous influence of this climate.

"That it has rendered them less able to generate an amount of animal heat sufficient to resist the intensity of the cold, while it has established a predisposition to the attacks of that disease (scurvy) the germs of which now so universally prevail amongst them, with its usual distressing influence on the mind, likewise rendering them highly susceptible of other diseases, and unable to withstand the privations to which they have been exposed, and which are inseparable from arctic service, is sufficiently obvious, as their present condition but too fully proves.

"It is likewise our opinion that, from their present state and condition, the remedial resources of the ship would be insufficient to establish such a state of health and efficiency as to afford any guarantee against the occurrence of those evils which could not fail to result from the circumstance of remaining in the ship, and being exposed to the intense severity of another (the fourth) arctic winter, after the effects of a sojourn so long as that which has fallen to their lot to have experienced.—We have, &c.

(Signed) "ALEXANDER ARMSTRONG,
Surgeon, H.M.S. Investigator.

(Signed) "W. T. DOMVILLE,
Surgeon, H.M.S. Resolute.

"HENRY KELLETT, Captain, C.B.,
H.M.S. Resolute."

Some days afterwards, even Captain M'Clure resigned himself to his disappointment, and, bitter as it was to be obliged to quit his ship, to look upon it as a duty. In every stage of his perilous voyage, he had found that all things were ordered for the best; and he dreaded lest the execution of what he and his officers considered their duty as naval men, that of saving their ship, should be construed into a charge of wantonly perilling his crew. It only remained, therefore, for him to land his boats, stores, and provisions, so as to form a depot for any one who, in after years, should need such supplies, and then to secure the Investigator to prevent her being blown to sea by future gales of wind.

As an instance of how carefully the provisions and stores had been husbanded during an absence of *three years* from any store or dockyard, the depot formed in the Bay of Mercy deserves a place in this narrative; it consisted of the following articles:—

Biscuit, ...	1000 lb.
Rum, ...	26 gall.
Brandy, ...	20 ,,
Salt beef, ...	600 lb.
Pork, ...	1600 ,,
Preserved meat, ...	3000 ,,
Flour, ...	6420 ,,
Suet, ...	112 ,,
Sugar, ...	1000 ,,
Chocolate, ...	435 ,,
Tea, ...	126 ,,
Tobacco, ...	484 ,,

Clothing for thirty men for a year, boat, spars, rope, powder, shot, and arms.

Their last duty, and that a painful one, was to erect a neat tablet over the graves of the three shipmates who had died in the Bay of Mercy; and that done, on the 3d June 1853, the Investigator's crew hoisted the colours to the mast-head of their dear gallant bark, and turned their backs upon her as sorrowfully as they would have done on an old well-tried friend in his extremity.

On June 17th, the squadron at Melville Island was reached by the Investigators, who found that every preparation for their comfort had been made on board the Resolute and Intrepid. Each ship housed a portion of the Investigator's crew and officers; and after a hearty meal and a long exchange of news from home and startling anecdotes from Banks Land, the Investigators settled into their new ships, and thought all their troubles at an end.*

Little occurred to break the monotony of arctic life at Bridport Inlet, Dealy Island, until Mr Roche, mate of the Resolute, who had been sent down early in the season to the North Star with some men, to relieve the increased demand upon the Resolute's provisions and stores, returned quite unexpectedly with a dog-sledge, having been to Beechey Island and back, a distance of 600 miles, within the short space of six weeks.

The news and letters he brought, together with the return of one travelling party after another of Captain Kellett's sledge expeditions, served to while away the anxious time before the water made in Barrow Strait.

Commander M'Clintock, Lieutenants Mecham and Hamilton, made extensive journeys, searching most completely every foot of land that lay to the north or

* See account of Mr Krabbé's visit to the Investigator in 1854, p. 263.

north-west of Melville Island. Commander M'Clintock's journey embraced the extraordinary distance of 1210 geographical miles, the longest ever made on foot in those regions on any one occasion; and Commander George Richards, by an equally remarkable journey from Northumberland Sound in Wellington Channel to the southern shores of Melville Island, informed Captain Kellett of the position of the Assistance and Pioneer.

Meantime Lieutenant Cresswell started for Beechey Island, with despatches to inform the Admiralty of the safety of the Investigator, in case the Intrepid and Resolute failed to escape from their advanced position in the forthcoming autumn.

When Captain Kellett had received the results of the search made by his sledge-parties, and saw that nothing farther could be done in a *north-west* direction, he decided to start with both vessels as soon as the state of the ice would admit of it towards Baffin Bay, and return at least as far as Beechey Island, the general rendezvous of all the searching expeditions.

August came in; the ice was in motion outside, but things did not look promising. The season was a backward one at Melville Island, though very forward in Wellington Channel. On the 18th of August a strong gale commenced off the land; the ice acted upon by it broke up in all directions, a lane of water made astern, and that night the Resolute and Intrepid were at sea, and the Investigators, poor fellows, as they hoped, homeward bound!

Within twenty-four hours the ships were brought up by the pack of Byam Martin Channel, and for many a day they lay under the extreme point of Melville Island, watching for an opening to dash across to Bathurst Land, knowing well that, once under its lee, northerly gales would inevitably make "*land water*," and enable them to accomplish another run for Beechey Island. Thence to England had now become a certainty.

Day after day passed; the drifting pack in Byam Martin Channel continued in a most unpromising state, whilst winter was fast advancing with snow, darkness, and newly-formed ice. Happily this part of Melville Island, like every other part of the southern shore of that favoured land, was found to be abounding in game, especially musk-oxen. Such a godsend, under the circumstances, was eagerly seized by Captain Kellett, who naturally felt most anxious to save and carry to England, in health and strength, the crew of the Investigator. All available guns and men were sent to secure fresh meat; and such was their success that about 10,000 lb. weight of game was eventually secured—and being soon frozen, it was easily preserved for the coming winter. At one time the meat was festooned round the rigging of the Resolute and Intrepid until they resembled butchers' stalls far more than British discovery-ships. At last, driven to risk anything rather than remain where they were for another winter, the vessels attempted to force a way through the pack; but on the 9th of September both the Intrepid and Resolute became permanently imbedded in the newly-formed ice, and a north-west gale forcing down the pack upon them, they became fairly beset and obliged to go whither it and Providence listed.

It was another disappointment to the gallant crew of

the Investigator. They met it with resignation, and a feeling of thankfulness that they were at any rate some 300 miles nearer home, and that in such well-found ships they would assuredly be carried in safety through their fourth winter. Indeed, no pains were spared by the officers and crew of the Resolute and Intrepid to grant every comfort to their passengers, and to distract their thoughts from those corroding anxieties which, perhaps, more than all else predispose to scurvy.

For two months the perils encountered by the drifting ships were very great. Their safety at last appeared to be occasioned by the body of heavy ice formed by constant pressure against the unyielding ships, the strength of which set at defiance the rest of the surrounding pack. At one time, with northerly winds, they feared being set down to the southward; and if there had been a good outlet for the ice between Lieutenant Osborn's and Lieutenant Wynniatt's farthest points in 1851, it was within the bounds of probability that next season, 1854, would have found the Resolute and Intrepid in some awkward position between Prince of Wales and Prince Albert Lands. This fear was put an end to when they found that the pack only drifted for a short time to the southward, as if to fill up tightly the great space called Melville Sound, and then it and the ships drifted steadily away to the eastward, recovering in some measure the southing that had been made, until the pack, doubtless checked by the islands which lay across its path towards Barrow Strait—such as those of Griffiths, Lowther, and Garrett—became stationary; and right pleased was Captain Kellett to find that after the 12th of November his good ship was at rest, and had then reached a point about due east of Winter Harbour, Melville Island, and in long. 101° W.—an admirable position for an early escape in the ensuing season. My duty, however, is to relate the adventures of the crew of the Investigator. Another winter passed over their heads without any great increase of disease. One officer, Mr Sainsbury (mate), died the 14th November, but he had been a very long time ill, and life was evidently prolonged so long as he continued to be buoyed up with the hope of escaping another winter in the ice.

Only two or three of the Investigators escaped this their last ordeal. Amongst them was Lieutenant Gurney Cresswell, who had been sent down to Beechey Island with a sledge-party in the spring; and the Phœnix, Captain Inglefield, having made a flying summer visit to that spot, Lieutenant Gurney Cresswell was taken to England, and on the 7th October 1853 communicated to the Admiralty the proud intelligence of the achievement of a North-West Passage, and the safety of Captain M'Clure and his companions.

To the wonderful supply of musk-ox beef obtained by Captain Kellett at Melville Island, the health of the Investigators was no doubt in a great measure due; but for that providential resource, the Resolute and Intrepid would have been able to afford them but a small supply of preserved meats, owing to the small quantity laid in when in England; and Sir Edward Belcher had moreover carried off the lion's share when we replenished from the North Star in August 1852.

As early as April 14, 1854, Captain M'Clure and his

crew quitted the Resolute and Intrepid, and proceeded on foot to Beechey Island—a pretty good proof that their health had not deteriorated since quitting the Bay of Mercy twelve months previously. One seaman alone was too feeble to walk; he was suffering from scurvy of the worst type, and soon after succumbed to it. The North Star, the depot-ship at Beechey Island, now became the spring resting-place of our friends the Investigators, and they there patiently awaited the means of escape to England.

That escape depended now upon the senior officer in those seas, Captain Sir E. Belcher. He, in the summer of 1853, before Captain Kellett left Melville Island, had started back to Beechey Island from the northern entrance of the Queen's Channel. The return sledge-parties of Commanders Richards and Osborn from Melville Island had told Captain Belcher of the position of the Investigator, and the accomplishment of a North-West Passage. To intercept the Resolute or Intrepid, if they touched at Beechey Island, appeared to be the object of Sir Edward Belcher. No time was, therefore, to be lost in opening a communication with Beechey Island; and so important was this deemed, that further search was abandoned, and one sledge-party was left to secure a retreat as best it could after a long and trying journey. Captain Belcher reached Cape Majendie at an early day; and there a boat, with Commander Pullen, from the North Star, joined the Assistance and Pioneer.

From Commander Pullen information was gleaned, that the season in Barrow Strait did not appear a forward one, and that most water was making on the west side of Wellington Channel; indeed, his heavy boat was a good proof that water and not ice had been mainly met with on his journey. Giving him orders that would prevent the Intrepid passing Beechey Island, and going to England direct with the news, as Kellett intended,* Captain Belcher sent Commander Pullen back again as quickly as possible, and, fancying he had provided for all contingencies, he commenced surveying that great curve of Wellington Channel called Baring Bay, and across which curve a solid floe was still firmly frozen. Any time between the 9th and 18th of August, the Pioneer could have towed the Assistance from Dundas Island direct to Cape de Haven; and indeed we are sure of it, from a trip made by Captain Inglefield in a boat to that cape from Beechey Island. The log-books of the Pioneer and North Star, and Inglefield's narrative, collectively attest the interesting fact, that water would have then been found by the Pioneer and Assistance.

No one was subsequently surprised when the Pioneer and Assistance were caught by the drifting pack, and beset at a place called Cape Osborn, 50 miles north of Beechey Island; though all were amazed when in the following early spring formal arrangements were made for the abandonment of all of H.M. ships within the arctic regions in 1854.

Totally ignorant of such a proceeding being the intention of the senior officer, the resources of the Resolute and Intrepid had been so carefully and judiciously hus-

* Captain Inglefield, however, in the Phœnix, arrived at Beechey Island, and carried home Lieutenant Cresswell with the despatches of Captain M'Clure.

banded, that with a reduced crew in each ship, they were still ready to meet the chance of not escaping in 1854; and this was the more creditable to Captains Kellett and M'Clintock, as they had had to victual the additional men and officers from the Investigator, and had left an ample depot of provisions and clothing in Melville Island for the use of Collinson, should fate lead him there.

Ever alive to the necessity of not ceasing to search so long as he was in a position to do so, Captain Kellett despatched in the early spring Lieutenant Mecham, supported by a party under Mr Krabbé (master), to revisit the Bay of Mercy in Banks Land, and to place on Princess Royal Island, in Prince of Wales Strait, information of the safety of the crew of the Investigator—a bold and happy act of foresight, as the result proved. Besides this, it was the intention of Captain Kellett to have sent parties later in the spring to connect Lieutenants Osborn's and Wynniatt's extreme points of search, and furthermore to have examined down Peel Sound.

Mr Krabbé reached the Investigator on 5th May. He says:—"The ensign and pendant were still flying. A large accumulation of snow-drift on her *north* side enabled him to walk in over her waist. On opening the fore-hatch, and going below, everything was found in good order, except that the ship had leaked so much as to be full with water up to her orlop-deck. The ship appeared to have dragged her anchors since she had been abandoned, for instead of being in nine fathoms water, she was then in eleven fathoms. This might have been occasioned by the movement of the ice, for there was no appearance of open water having existed in Mercy Bay in the autumn of 1853." Mr Krabbé finally left her on May 11, 1854.

In the mean time Captain Richards—who was despatched in weather so severe as to endanger the lives of all his party—reached Captain Kellett with a "confidential" letter from Sir Edward Belcher. That "confidential" letter is, of course, now a public document, and a very remarkable one too. It contains this paragraph, which is here copied verbatim:—

"Should Captain Collinson fortunately reach you, you will pursue the same course, and not under *any consideration risk* the detention of *another season*. These are the views of the Government; and having so far explained myself, I will not hamper you with farther instructions than, Meet me at Beechey Island, with the crews of all vessels, before the 26th August."

Captain Kellett determined not to adopt any such course upon a *confidential letter*, and immediately despatched Captain M'Clintock to Sir Edward Belcher, to point out the perfect feasibility of saving his ships—to assure him of the provisions and stores, as well as the health of a sufficient number of officers and men, being such as would enable him to meet the possible contingency of another winter, rather than abandon her Majesty's ships, when they lay in the very best position for an escape directly the ice broke up in Barrow Strait, and finally, to point out to Sir E. Belcher that he was strongly against the desertion of so many fine ships. But the representations of Captain Kellett were unavailing. Captain Belcher sent Captain M'Clintock back with an *order* for the abandonment of the Resolute* and Intrepid; and the crew of the Investigator, who had lived through such trials and hardships for four winters, stared to see all hands gradually retreating upon Beechey Island, ready to return to England.

Lieutenant Mecham and Mr Krabbé's parties returned during the summer to Beechey Island, having performed marvels in way of the distances they traversed. Lieutenant Mecham brought from the Investigator's depot of provisions on Princess-Royal Island deeply interesting information touching the movements of H.M.S. Enterprise, dated August 1852. This was the first news any one in Barrow Strait had had of the Enterprise since she passed Behring Strait in 1851. Captain Collinson distinctly said, "It is my intention to pursue the channel separating Wollaston from Prince Albert Land, the entrance to which is in lat. 70° 30′ N.;" we all therefore naturally supposed that some volunteers would be left behind to help him, in case he should be imprisoned in some ice-bound harbour south-west of Cape Walker. But Sir Edward Belcher got rid of all difficulty as to Captain Collinson's safety by the following train of reasoning. He says:—

"However anxious I may be for a similar result† with regard to Captain Collinson and party, still I am thankful that *the records place him in a region free from the perils of arctic ice*, in which Captain M'Clure con-

* The Resolute has since been picked up by an American whaler, drifting out into the Atlantic; and so little had she suffered in her lonely vyoage that the paint-work was not even scratched by the ice. Thanks to the Americans, she is again under an English flag.

† He refers to the safety of the Investigator's officers and men.

siders no ship could endure. He had at the latest account two modes of escape: one, by the road he came; the other, on which I place but little reliance on account of its difficulties, by the land journey to some of the Hudson Bay posts; unless, indeed, he met Dr Rae, in which case competent guides would materially alter the face of his difficulties."

With these ambiguous terms he left Captain Collinson to get out of his difficulties as best he might, and next proceeded to give an appearance of intending to stand by and save the Assistance and Pioneer. They on the 6th August broke out of winter-quarters, and advanced slowly down channel as the water and pressure from the north began slowly to break up the belt of ice which extended across Wellington Channel.

The ice in Barrow Strait broke up at the same time; and by the 22d August the floe of Wellington Channel had dissolved to a distance of fifteen miles northward from Barrow Strait. A belt of ice, about twenty miles wide, only remained between the ships and the waters communicating with the Atlantic Ocean, and that belt of ice much cracked, and evidently working with every tide; yet it was determined to quit them, and on the 26th August 1854 the last of that ill-fated expedition was deserted.* All the officers and men of H.M.S. Assistance, Resolute, Pioneer, Intrepid, and Investigator had just got on board the North Star and made sail, when the Phœnix, Captain Inglefield, and Talbot, store-

* In the season of 1854-55, two vessels are said to have fallen into the hands of the Esquimaux of Ponds Bay; in all probability the Resistance and Pioneer drifted there.

ship, Commander Jenkins, hove in sight round Cape Riley.

A fresh division of the men and officers then took place, to relieve the crowded decks of the North Star, and the Phœnix and Talbot each received a portion. That and some other small arrangements completed, no time was lost in beating a retreat; and on September 6th all reached Disco, on the coast of Greenland, in safety, and eventually returned to England September 28, 1854.

The gallant Investigators found all England's sympathies and feelings enlisted in the war with Russia; and although the members of a naval court-martial, which went through the established form of inquiring into the loss of H.M.S. Investigator, most honourably acquitted Captain M'Clure, his officers, and men, from any blame on her account, and added the highest encomiums upon the gallantry and zeal exhibited by all, yet, in a public point of view, the ship's company generally felt that few honours were awarded to them in comparison with the sufferings they had so nobly borne. The Admiralty, to evince their approval, dated back Captain M'Clure's commission, as well as that of his first-lieutenant and some other officers, to the day on which the North-West Passage was discovered. Her Most Gracious Majesty shortly afterwards conferred the honour of knighthood upon Captain M'Clure; and assuredly it never was more worthily bestowed.

In the following session of Parliament a select committee of the House of Commons met, to take into consideration the reward due to those who had discovered and achieved the North-West Passage; but in the interim between the arrival of Captain Sir Robert M'Clure in England and the meeting of Parliament, news had arrived that Dr Rae had obtained certain information of a party from Franklin's missing squadron having passed the intervening unknown space which lay between Barrow Strait and the coast of North America. The duty of the committee became a somewhat more responsible one, in so far as it had to award the priority of discovery to Franklin or M'Clure, before the papers of the former came to hand.

Lady Franklin, in a most able and touching letter, called the earnest attention of the honourable committee to the impossibility of arriving at any certain decision in the absence of all evidence as to Franklin's claim to the priority; and they therefore qualified the award by stating, very justly, that Captain Sir Robert M'Clure, in H.M.S. Investigator, had discovered *a* north-west passage, and successfully carried his followers from the Pacific to the Atlantic Ocean by that route, exhibiting himself an example of unflinching perseverance, courage, and zeal, which his officers and men nobly followed, and, to use the words of the honourable committee, "that they performed deeds of heroism which, though not accompanied by the excitement and the glory of the battle-field, yet rival, in bravery and devotion to duty, the highest and most successful achievements of war!" Accordingly a reward of £10,000 was granted to the officers and crew of H.M.S. Investigator, as a token of national approbation; and, acting upon a suggestion thrown out by the honourable committee, all this gallant ship's company eventually received from their Queen a medal, which they will assuredly treasure far more than any pecuniary reward.

CHAPTER XIX.

GLOOMY PROSPECT IN THE AUTUMN OF 1854—REVIVAL OF DESPONDING TALES—SUDDEN ARRIVAL OF INTELLIGENCE FROM DR RAE—A PARTY FROM FRANKLIN'S SHIP HEARD OF—DR RAE'S REPORT—RELICS AND PROOFS OF BOTH EREBUS AND TERROR BEING IN EXISTENCE—THE RUSSIAN WAR PREVENTS A NAVAL EXPEDITION BEING SENT TO BARROW STRAIT—THE ADMIRALTY DIRECT THE HUDSON BAY COMPANY TO SEND MR ANDERSON—MR ANDERSON PROCEEDS IN 1855 TO THE MOUTH OF THE GREAT FISH RIVER—VERIFIES THE FACT OF A PARTY FROM THE LOST EXPEDITION HAVING BEEN THERE—NO LIGHT THROWN UPON THEIR FATE—NEITHER BODIES, GRAVES, CLOTHING, NOR ARMS DISCOVERED—REMARKS UPON THE RELICS DISCOVERED—PROBABLE COURSE ADOPTED BY THE FORLORN HOPE—HOPES EXIST OF THE MYSTERY STILL BEING CLEARED UP—DISTANCE THE PARTY COULD HAVE TRAVELLED—POSITION OF THE LOST EXPEDITION: HOW LOST—REASON WHY FURY BEACH WAS NOT VISITED BY THEM—CREDITABLE TO ENGLAND THAT THE SEARCH HAS NEVER BEEN STAYED—THE ADMIRALTY REWARD DR RAE FOR GIVING US INFORMATION OF FRANKLIN'S POSITION—GENERAL REVIVAL OF INTEREST IN THE QUESTION.

NOTHING could have been more crushing to the hopeful feelings of even the most sanguine or earnest in the search for Franklin than the sad intelligence which was brought home in October 1854. The labours of the officers and crews of H.M.S. Assistance, Resolute, Intrepid, and Pioneer, during three years, had not thrown one ray of light on the sad mystery! and as if to preclude all possibility of any further steps being taken, all the vessels available for the search had been wantonly abandoned, deserted to the mercy of the ice and winds of 74° north latitude. The solemn silence with which the venerable president of the court-martial which sat to try Captain Belcher returned him his sword, with a bare acquittal, best conveyed the painful feelings which wrung the hearts of all professional men upon that occasion; and all felt that there was no hope of the mystery of Franklin's fate being cleared up in our time, except by some unexpected miracle. But just at that very time, when those who had ever taken a gloomy view of the subject smiled at the realisation of their unfavourable prognostications, and congratulated themselves on having exactly foretold what had happened to the Erebus and Terror (either hoisting them up on the top of floes off Newfoundland or squeezing them to destruction in Lancaster Sound), a letter reached England from Dr Rae, announcing that he had at last struck upon the clue, and that a portion of Franklin's expedition had reached and perished at or near the mouth of the Great Fish River!

Is it presumption to say that the opportune discovery of such a fact at such a moment was a marked instance of Divine interposition? That it should have come from such a quarter is all the more interesting, because Dr Rae, whilst on the journey in which he became possessed of this important information, was, he tells us, purely employed upon geographical research; and prior to starting he announced that fact, coupling it with the remark that he was going where Franklin was not likely to be met with.

His tale is briefly this. He had been sent by the Hudson Bay Company in 1853 to complete the survey of the long isthmus of land which connects North Somerset with the American continent under the name of Boothia. He had to connect Captain Sir James Clarke Ross's magnetic pole, or the coast-line about it, with his own discoveries near the Castor and Pollux River.

Repeating his old plan of proceedings in 1846-47, Dr Rae wintered at the lakes on the isthmus which divides Regent Inlet from Repulse Bay, and early in the spring of 1854 started with his sledge-party to accomplish his task. Ascending Committee Bay as far as Simpson Peninsula, he then struck westward, taking advantage of a series of lakes and frozen streams to relieve the labour of sledging across the land which intervened between him and the western waters. Whilst making his way in that direction, Dr Rae met, on the 20th April, an Esquimaux, who, upon being asked if he had ever seen any ships or white men, replied no, but that a party of white men had died of starvation a long distance to the west of where he then was, and beyond a large river!

Now distance and time are two things of which an Esquimaux has very vague ideas; and Dr Rae assures us that, although he afterwards had reason to believe that the Great Fish River, then only seventy or eighty miles distant, was the stream referred to, he could only learn that the spot spoken of was beyond a distant river. Unable to glean more particulars, further than here and there coming across convincing proofs that the natives were in possession of articles from Franklin's ships, Dr Rae made an effort northward, as if to combine an execution of his instructions with the purpose (as he assured the writer) of proceeding in the direction the retreating party must have taken when coming down upon the American shore. Circumstances prevented his journey being successful. Dr Rae returned to the mouth of the Castor and Pollux River, and again retraced his steps overland to Repulse Bay, picking up relics and information, the result of which he condensed into the following Report to the Lords Commissioners of the Admiralty, and then made the best of his way to England:—

"On the morning of the 20th (April) we were met by a very intelligent Esquimaux, driving a dog-sledge laden with musk-ox beef. This man at once consented to accompany us two days' journey, and in a few minutes had deposited his load on the snow, and was ready to join us. Having explained to him my object, he said that the road by which he had come was the best for us; and having lightened the sledges, we travelled with more facility. We were now joined by another of the natives, who had been absent seal-hunting yesterday, but, being anxious to see us, had visited our snow-house early this morning, and then followed up our track. This man was very communicative; and on putting to him the usual questions as to his having seen 'white men' before, or any ships or boats, he replied in the negative, but said that a party of 'Kabloonans' had died of starvation a long distance to the west of where we then were, and beyond a large river. He stated that he did not know the exact place, that he never had been there, and that he could not accompany us so far. The substance of the information then and subsequently obtained from various sources was to the following effect:—

"In the spring, four winters since (1850), while some Esquimaux families were killing seals near the north shore of a large island, named in Arrowsmith's charts King William Land, about forty white men were seen travelling in company southward over the ice, and dragging a boat and sledges with them. They were passing along the west shore of the above-named island. None of the party could speak the Esquimaux language so well as to be understood; but by signs the natives were led to believe that the ship or ships had been crushed by ice, and that they were now going to where they expected to find deer to shoot. From the appearance of the men, all of whom, with the exception of an officer, were hauling on the drag-ropes of the sledge, and looked thin, they were then supposed to be getting short of provisions; and they purchased a small seal, or piece of seal, from the natives. The officer was described as being a tall, stout, middle-aged man. When their day's journey terminated, they pitched tents to rest in.

"At a later date the same season, but previous to the disruption of the ice, the corpses of some thirty persons, and some graves, were discovered on the continent, and five dead bodies on an island near it, about a long day's journey to the N.W. of the mouth of a large stream, which can be no other than Back's Great Fish River (named by the Esquimaux Oot-koo-hi-ca-lik), as its description and that of the low shore in the neighbourhood of Point Ogle and Montreal Island agree exactly with that of Sir George Back. Some of the bodies were in a tent, or tents; others were under the boat, which had been turned over to form a shelter, and some lay scattered about in different directions. Of those seen on the island, it was supposed that one was that of an officer (chief), as he had a telescope strapped over his shoulders, and a double-barrelled gun lay underneath him.

"From the mutilated state of many of the bodies, and the contents of the kettles, it is evident that our wretched countrymen had been driven to the dread alternative of cannibalism as a means of sustaining life. A few of the unfortunate men must have survived until the arrival of the wild fowl (say until the end of May), as shots were heard and fresh bones and feathers of geese were noticed near the scene of the sad event.

"There appears to have been an abundant store of ammunition, as the gunpowder was emptied by the natives in a heap on the ground out of the kegs or cases containing it, and a quantity of shot and ball was found below high-water mark, having probably been left on the ice close to the beach before the spring commenced. There must have been a number of telescopes, guns (several of them double-barrelled), watches, compasses, &c., all of which seem to have been broken up, as I saw pieces of these different articles with the natives, and I purchased as many as possible, together with some silver spoons and forks, an Order of Merit in the form of a star, and a small silver plate engraved 'Sir John Franklin, K.C.B.'

"Enclosed is a list of the principal articles bought, with a note of the initials, and a rough pen-and-ink sketch of the crests on the forks and spoons. The articles themselves I shall have the honour of handing over to you on my arrival in London.

"None of the Esquimaux with whom I had communication saw the 'white' men, either when living or after death; nor had they ever been at the place where the corpses were found, but had their information from those who had been there, and who had seen the party when travelling on the ice.

"From what I could learn, there is no reason to suspect that any violence had been offered to the sufferers by the natives.

"List of articles purchased from the Esquimaux, said to have been obtained at the place where the bodies of the persons were found, viz.:—

"One silver table-fork—crest, an animal's head, with wings extended above; 3 silver table-forks—crest, a bird with wings extended; 1 silver table-spoon—crest, with initials 'F. R. M. C.' (Captain Crozier, Terror); 1 silver spoon and one fork—crest, bird with laurel branch in mouth, motto, '*Spero meliora;*' 1 silver table-spoon, 1 tea-spoon, and one dessert-fork—crest, a fish's head looking upwards, with laurel branches on each side; 1 silver table-fork—initials, 'H. D. S. G.' (Harry D. S. Goodsir, Assistant-Surgeon, Erebus); 1 silver table-fork —initials, 'A. M'D.' (Alexander M'Donald, Assistant-

Surgeon, Terror); 1 silver table-fork—initials, 'G. A. M.' (Gillies A. Macbean, Second-Master, Terror); 1 silver table-fork—initials, 'J. T.'; 1 silver dessert-spoon—initials, 'J. S. P.' (John S. Peddie, Surgeon, Erebus); 1 round silver plate, engraved, 'Sir John Franklin, K.C.B.;' a star or order, with motto, '*Nec aspera terrent*, G. R. III., MDCCCXV.'

"Also a number of other articles with no marks by which they could be recognised, but which will be handed over, with those above named, to the Secretary of the Hudson Bay Company.

"JOHN RAE, C.F.

"REPULSE BAY, *July* 1854."

It matters little what portion of the Esquimaux tale is correct, or what fabulous; of one great fact Dr Rae assured us, that a party from the Erebus and Terror did reach the coast on or about the Great Fish River. It is fair to infer that the party comprised officers and men from *both* vessels, because the few articles recovered from the natives bear the names of Erebus as well as Terror. For instance, we have Franklin's star of the Guelphic order, and some of Crozier's plate.

It was very natural,—for it occurred in Captain M'Clure's case, as told in the foregoing narrative,—that if the ships of Franklin's expedition had become frozen in in some bay which did not often clear out of ice, he should have done as M'Clure intended to do—send a party home *viâ* America, to convey intelligence and seek succour. Franklin knew that when, on a former occasion, in 1833, Sir John Ross had got into difficulties in the Victory, and was missing, a party was at once organised, and sent down the Great Fish River to seek for him. Might not Sir John Franklin have fairly supposed that as much would be done in his behalf? How could he know of the opposition all propositions of such a rational nature were likely to meet with from persons consulted by the Admiralty?

The public mind was too deeply engaged in the sufferings of the British army upon the heights of Sebastopol, to grant the attention it merited to the interesting intelligence brought to England by Dr Rae, or to the collateral proof brought home by Captain Collinson (who had happily escaped with the Enterprise) from another point about the same distance from the north shore of King William Land as the Great Fish River. It consisted of a piece of wood-work, which must have belonged to either the Erebus or the Terror, and which was found by accident upon an island near the Enterprise's winter-quarters in Cambridge Bay.

The Lords Commissioners of the Admiralty took the opinion of some arctic authorities upon the subject of what could be done towards still further clearing up the tale brought home by Dr Rae; for there was much about it that was vague, and calculated to keep alive hopes of the most distressing nature to those deeply interested in the crews of Franklin's ships. A gigantic war was pressing upon the resources of our navy both in ships and men,—none of them could then be spared; and to meet the outcry of some effort to be made to ascertain if it really was the mouth of the Great Fish River that Franklin's travellers had reached, the Hudson Bay

Company were again requested to send out a party to that locality.

Dr Rae having declined to take charge of the party which was equipped for this purpose, though he gave every support and encouragement to it, it was consigned to the care of Mr James Anderson, a chief factor of the company, an officer of high reputation and much experience as a traveller. Lady Franklin, however, earnestly protested against this expedition; she foretold the improbability of its ever reaching King William Land, and short of that the result would be as inconclusive as Dr Rae's report, and a loss of very valuable time.

Labouring under many disadvantages, from the short time given to equip and start, Mr Anderson commenced his descent from Fort Resolution to the mouth of the Great Fish River, on June 22, 1855, with three canoes of wooden framing, but birch-bark planking, *without an Esquimaux interpreter*. On July 30, at the rapids below Lake Franklin, three Esquimaux lodges were seen, and various articles were found, denoting that some of the unfortunate men they were in search of had been there. The foot-note to this page is worthy of careful perusal—we will refer to it again.*

Pushing on, Point Beaufort was reached, and at last Montreal Island landed upon. "There," says Mr Anderson, "on a high ridge of rocks at the S.E. point of the

* Extract from Mr Anderson's Report, *vide* Blue-Book:—"On the 30th, at the rapids below Lake Franklin, three Esquimaux lodges were seen on the opposite shore, and shortly after an elderly man crossed to us. After the portage was made we crossed over, and immediately perceived various articles belonging to a boat, such as tent-poles and kayack paddles made out of ash oars, pieces of maho-

island, a number of Esquimaux *caches* were found, and, besides seal-oil, various articles were found belonging to a boat or ship, such as chain-hooks, chisels, blacksmith's shovel and cold chisel, tin oval boiler, a bar of unwrought iron, about three feet long, one and a half inch broad, and a quarter of an inch thick; small pieces of rope, bunting, and a number of sticks strung together, on one of which was cut 'Mr Stanley' (surgeon of the Erebus). A little lower down was a large quantity of chips, shavings, and ends of plank of pine, elm, ash, oak, and mahogany, evidently sawn by unskilful hands; every chip was turned over, and on one of them was found the word 'Terror' carved. It was evident that this was the spot where the boat was cut up by the Esquimaux; but not even a scrap of paper could be discovered, and though rewards were offered, and the most minute search made over the whole island, not a vestige of the remains of our unfortunate countrymen could be discovered."

The party next examined Point Ogle, where only a small piece of cod-line and a strip of cotton were found; and on the 8th August they began to retrace their steps, having held no communication with, indeed seen, no Esquimaux beyond the one man and few women at the rapids below Franklin Lake, and never been able to reach King William Land. This information reached us early in 1856, and went to confirm Dr Rae's supposition

gany, elm, oak, and pine; also copper and sheet-iron boilers, tin soup-tureens, pieces of instruments, a letter-nip with the date 1843, a broken hand-saw, chisels, &c. Only one man was left at the lodges; but the women, who were very intelligent, made us understand, by words and signs, that these articles came from a boat, and that the white men belonging to it had died of starvation."

that the Great Fish River was the stream upon which the party he had heard of had retreated; but instead of clearing up the mystery of what had become of them, the whole story left the fate of Franklin, Crozier, and their ships' companies as doubtful as ever.

Taking it for granted that the Esquimaux did see thirty or forty men with a boat, as Dr Rae asserts, what had become of them? If, when they reached the continent, the unfortunates became desperate with misery and committed cannibalism—the practice is by no means rare in those wild regions, and it would assuredly prolong life—where were the survivors? Is it likely they sat down there and died one after the other? If they were so lost to their own safety as to remain, would not the survivors have scraped the earth over the bones of those who first perished?

Every arctic traveller knows that the tender and oily bones of the seal—even the brittle ones of birds—are found preserved over the whole extent of the arctic regions visited by us. What, then, had become of the bones of thirty men? Five years after the Erebus and Terror left Beechey Island, in Barrow Strait, all those who visited the scene of their winter-quarters found *clothing, scraps of paper*, and the thousand signs of Europeans having been there, looking just as fresh as the day they were left, and that in a far worse climate than Montreal Island.

Thirty-one years after Sir Edward Parry had been at Bushnan Cove, Melville Island, a traveller (Lieutenant M'Clintock) found a spot where that distinguished navigator had, to use his own words, made "a sumptuous meal of ptarmigan," and there lay the bones of those very birds strewed about the old encampment! "I was astonished," says Lieutenant M'Clintock (*vide* Parliamentary Blue-Book, 1852), "at the fresh appearance of the bones; they were not decayed, but merely bleached, and snapped like the bones of a bird recently killed."

Esquimaux were not likely to have used dead men's bones. If they had European clothing in their possession, it is hardly likely that they could have concealed it entirely. There was not a musket, pike, or cutlass produced. The party were not likely to have gone there unarmed; indeed the Esquimaux acknowledged having seen both powder and shot and ball. And as to Mr Anderson's theory of the wind blowing away or covering their journals and papers, because his nautical almanacs suffered, it is purely assuming that the officer who headed Franklin's party was such an idiot as to leave his papers strewed about the surface of Montreal Island, instead of putting them in a *cache*, where, as arctic discovery proves, papers have been preserved and discovered after longer intervals of time than perhaps any other climate would admit of.

Looking, therefore, at the evidence adduced, it amounted simply to this, that

"A party from the Erebus and Terror did reach the Great Fish River, and left traces at Montreal Island and at the first rapids in ascending the stream!" Further than this, all was apocryphal. Mr Anderson very naturally went upon his journey, firmly believing every iota of the translated account of Dr Rae's interpreter; indeed, in the absence of any means of communication with the one old man and few women whom he did see, he had no other resource than to connect the traces which lay before him with the report previously made public. But sailors may be allowed to put a sailor's explanation to what lay before Mr Anderson, and the following is our version of the tale it told:—

On Montreal Island Mr Anderson found, he says, "a quantity of chips, and *shavings,* and the ends of *plank* of pine, elm, ash, oak, and mahogany, evidently sawn by unskilful hands."

Now, no boat supplied to the Erebus or Terror from her Majesty's yards, which any party of men could have dragged a hundred miles over ice, would have been constructed of *plank* of so many descriptions; but it is very certain that a party retreating to the Great Fish River, and knowing the long series of rapids and portages in that stream, would have carried with them materials such as plank, which, with the framing of their large boat, would form rough canoes fit for their purpose.

Mr Anderson distinctly says "chips and *shavings.*" Now a savage, who had never seen a planing instrument, was not likely to be able to produce shavings. After informing us that the plank was evidently cut by unskilful hands, Mr Anderson says, "*Every chip was turned over, and on one of them was found the word 'Terror' carved!*" Surely that ominous word is a mute witness against Esquimaux having been the men who there laboured; yet in the next paragraph we read, "It was evident that this was the spot where the boat was cut up by the Esquimaux!"

Surely no such fair inference can be drawn. That the

party brought carpenter's tools with them, we have the proof in Mr Anderson discovering, at the lodges near the rapids, "*a broken hand-saw, chisels, &c.;*" and perhaps, if a careful list could be procured of every article seen there or at Repulse Bay, some more interesting evidence might be obtained; for even as a straw will show the course of a great stream, so may some insignificant trifle throw sudden light upon this sad subject.

The existence of traces further up the river than Montreal Island was a significant fact; and supports our idea that on Montreal Island preparations were made to ascend the stream, of which indeed we have another proof in the ash oars being cut or reduced into paddles,—a very necessary measure for a party about to go up narrow and tortuous rivers, and unlikely to have been done by the Esquimaux, who have no kyacks or canoes in that part of America. Some of these paddles were found at the rapids likewise.

It is true the women at this spot made signs that these articles came from a boat whose crew perished of starvation; but they did not give a single proof of the truth of the tale, or point out the grave of one of the unfortunate party.

Dr Rae, zealous for the character of the Esquimaux, repudiates indignantly all idea of their having been treacherous, nor is it at all desirable to give rise to any bloody suppositions upon the matter; but any one who will carefully read over the able paper of Captain Maguire, in the Appendix of this work, can, as easily as the most experienced traveller, form a correct idea of the character of the Esquimaux generally; and he will then agree with us in thinking that the savage of the polar regions, though not naturally cruel or treacherous, would, like most others, consult his own interests rather than the dictates of humanity, when such a windfall as a boat's crew of starving, scorbutic men, carrying with them untold wealth in the shape of wood, iron, and canvass, fell into their hands, and when they confessed, as those poor fellows evidently did, their direful necessity.

Some of Franklin's people may, we think, have died of disease or starvation at the place upon the continent spoken of by the natives; but that spot has not been reached by us as yet. Others evidently got to an island; there the Esquimaux say the officer perished, and five men likewise. Such an island as Montreal Island was very likely to have been chosen by our starving countrymen to await the opening up of the Great Fish River; they would be in a good position for commencing their canoe voyage, and be less likely, whilst employed constructing canoes or rafts, to be interrupted by natives. Granted, therefore, that some starved at each place spoken of by the natives—granted even that the remainder did so far forget their manhood as to eat the flesh of their shipmates,—is it unreasonable to suppose that, when the river opened, some few of those unfortunates started with what they had constructed, abandoning all their unnecessary gear on the island, and at the first portage?

They might have ascended far, and fallen in detail, and yet never, in such a water-intersected region, have been discovered by Mr Anderson in his descent—the more especially if they, taking Sir George Back's chart, had followed his old track, a track from which Mr Anderson departed considerably, and with advantage to himself and his party as far as rapidity of journey was concerned. As to holding out a hope of any straggler surviving amongst Esquimaux or Indians, it is not our desire to do so; but those who, by following up a similar train of argument as ourselves, arrive at a hope of such a pleasing and consolatory nature, ought not to be ridiculed for doing so.

They who have kept alive hope, who have urged on expedition after expedition, in spite of failure, in spite of ridicule, and in spite of uncharitable imputations of mania or interested motives, have now reason to feel happy that such trifles did not check their efforts; it remains yet to be seen whether perseverance will not still lift the curtain of this sad but glorious tragedy.

It is not alone the fate of those forty men that we desire to know—they were but a fraction of the lost expedition; there are still one hundred souls unaccounted for, and two of her Majesty's ships!

To those who urge the expense of arctic expeditions, or the risk of life, as objections to the completion of a task we are pledged to accomplish, the answer is a brief one. Read the long list of soldiers and sailors who yearly go to their unhonoured graves in the pestilential Bight of Benin, on the fevered shores of the Western Indies, the cholera-ravaged stations of Hindostan or China, where the charge for medical stores is sometimes greater than that for feeding the perishing thousands! For what do they lay down their lives? Is it because we desire that all nations may honour a people who will and can do all things and dare all things, because it is right? or do we merely weigh the sale of cottons and hardware against the lives of our brother men?

If the former is the rule, then let England dare to do right, and risk the charge of Quixotism, even though, in exploring the frozen zone, the extension of free-trade principles or the regeneration of the negro race may have no share. And for expense, let one line-of-battle ship the less be kept in commission until the question is settled, or some other retrenchment be made, if we are in such a bankrupt condition that England cannot afford to seek her missing sons.

We, who have reduced arctic travelling to a mere arithmetical calculation, know very nearly the distance a body of sailors numbering forty could have come from, dragging a heavy wooden boat over the ice, besides the quantity of articles which have been enumerated elsewhere, and which formed, doubtless, but a small portion of what they had with them. Taking, therefore, the weight dragged by the forty men as 200 lb. per man, and the distance accomplished daily about ten miles, an allowance extremely liberal for debilitated seamen, we have the precedents of Captains Richards, Osborn, and Penny (who all have had to carry heavy wooden boats as far as possible over the ice) for saying that a journey of about fifteen days, or 150 miles, would be about the utmost distance they could have come from; the more so that sledge-travelling was then but little understood, and that the extent of the sledge-journeys made from Beechey Island by Franklin's people do not exhibit any marked improvement.

That the Erebus and Terror are somewhere within the

limits of the unsearched area about King William Land, everything now denotes. One hundred and fifty or two hundred miles from Montreal Island, northward, carries us into the centre of this space, and where Victoria Strait is split in two by the large island called King William Land. In and about Cape Felix on that island, or near the magnetic pole in Boothia, they most probably got beset; for had they been on Victoria Land, where natives, game, and fish abound, they would, it is fair to infer, have sent their "forlorn hope" along it towards the Coppermine or Mackenzie River. How they reached that supposed point, with their ships, time and a discovery of their journals will alone tell. Whether by rounding the *west* side of Prince of Wales Land, and passing down a channel which some suppose to exist in a south-east direction between it and Victoria Land, or whether, as appears most natural, they took the fine and promising channel which offered to the southward between Cape Bunny and Cape Walker, now called Peel Sound, and so struck the American continent, we can only surmise. But the absence of all cairns, or signs of their having been detained or having landed on either coast of Prince of Wales Land, as far as it is now known, or of North Somerset, leads to the natural supposition that they are nearer to King William Land than to any other spot—perhaps in some indentation on its northern coast, into which they ran during a late and stormy season, as M'Clure did in the Investigator, and John Ross did in the Victory, never to escape with their ships.*

* The Editor can now appeal to the subsequent discovery of Franklin records and relics on the very spot here indicated, as a proof of the correctness of his views.

It has been argued against the existence of Franklin's ships in that quarter, that he would assuredly have visited the Fury Beach depot, in Regent Inlet. We reply to this, that Franklin, through his ice-master and others in his expedition, knew well how worthless it was for his purpose. He knew that, since it had been formed, Sir John Ross had provisioned the Victory from it, that he had retreated upon it, and lived on it with his crew nearly twelve months, and eventually equipped himself there prior to his escape in 1833. After that some whalers had swept nearly everything off the beach; and, to escape the consequences of an Admiralty prosecution, one of the vessels had thrown into Peterhead Harbour a quantity of provisions she had carried off as plunder from the Fury depot.

It would be unfair to discuss the question of who has been to blame for fruitless efforts, or to assert that the zeal and energy of officers has been fruitlessly expended in the search for Franklin up to 1854. It could not be otherwise. The chart as it stood in 1848 was a blank. The labours of those employed gradually narrowed the area to the mere work of one season; and to those who blame us for having spent time in searching to the north-west of Beechey Island, the simplest reply is, that we glory in having had the hope accorded to us in 1852 of Franklin's expedition being in that direction. It kept up the interest of the world upon the subject; and it enabled us, though unsuccessful there, to say we never desponded, and never believed that they would not be found, or that they turned back from Beechey Island; and England may boast that, owing to that and other circumstances, she never relaxed her efforts until a certain clue to their position was secured. It was undoubtedly for that clue, with the certainty that it held out of our soon knowing the fate of Franklin's and Crozier's ships' companies, that the Admiralty rewarded Dr Rae with a portion of the twenty thousand pounds awarded by Parliament; and although such a reward does not come under the strict interpretation of the Act, still there is no doubt of their Lordships having generously exercised their prerogative, in stretching the rule, and rewarding an active and zealous arctic traveller for obtaining a trace which was worth twice the sum, and which gave fresh hope and spirit to all who thought upon the subject.

The wonderful voyage and journey of the Investigator's gallant crew was about to be followed by the remarkable discovery, by Captain M'Clintock, of the touching record of Franklin's sufferings, and the fate of his noble followers. That discovery, the last great arctic achievement of our generation, fell, it is true, to a private expedition, and the honour, apart from its leader being a naval officer, cannot be claimed by our profession. The fault, however, of desisting from the search for Franklin, just as success was certain, arose from official ignorance of the subject, and the alarm created by Sir Edward Belcher's strange proceedings during the last expedition to Barrow Strait; and it is only fair to those whose hope flagged not, and whose energies were constantly directed to urging the final expedition upon the attention of the Admiralty and Government, to place on record their last appeal before they turned to private sources to procure the means of completing a task which red-tape despaired of.

The first petition emanated from her who has been the mainspring of the search, her to whose untiring energy in pushing forward fresh expeditions we owe the accomplishment by naval officers of the discovery of the North-West Passage, and then the perfect certainty of discovering the long lost and sought Erebus and Terror; the other was promoted by Sir Roderick Murchison, a steadfast advocate in every step taken to save the missing expedition; and, supported as he has been by the larger portion of the scientific men of our day, he and they have in no small degree contributed to the solution of many geographical and physical problems in our arctic zone, and assisted to raise the veil which once hid the fate of Franklin's expedition.

We cannot better take leave, for a while, of arctic discovery, and the no less honourable search for our countrymen, than by using the quaint but eloquent words of the right worthy Samuel Purchas, parson of St Martin's, by Ludgate, London.

"Great jewels are those merchants and mariners which, to the glory of our nation, neither spare cost, and fear no danger in these attempts—resolute, gallant, glorious attempts! which thus seek to tame nature where she is most unbridled, and to subdue her to that government and subjugation which God hath imposed on all things to the nature of man. Great God! let me in silence admire and worship thy wisdom, that in this little heart of man hast placed such greatness of spirit as the world is too little to fill."

LADY FRANKLIN'S LETTER.

"60 PALL MALL, 11*th July* 1856.

"MY LORDS,—Three months ago I felt constrained to address a letter to your Lordships, requesting that you would be pleased to delay your adjudication of the reward claimed by Dr Rae for ascertaining the fate of my husband's expedition until such time as the result of a more complete and final search could be known. I implored your Lordships to adopt such measures as would set this question at rest, and, at the same time, was compelled to represent that your refusal to do this would force upon me the painful alternative of taking the burden of an expedition upon myself, at whatever cost, and under great disadvantage.

"To this letter I have not been honoured with any reply; but, notwithstanding, it seemed to me, and to others, not unreasonable to interpret your silence in a manner not unfavourable to my wishes, inasmuch as your Lordships were well aware that so long as no adverse decision was announced to me, I was precluded from taking any steps for advancing my private expedition, which depended entirely on the non-adoption of the other. Even when I read in the 'Gazette,' after two months and more had elapsed, that your Lordships, disregarding my request, had given the reward of £10,000 to Dr Rae, I was still unwilling to regard this act as an absolute rejection of my petition for further search, since in that light, or with such an object in view, it might have been practicable to announce it at a much earlier period, and thus relieve me from suspense, and set my hands free for action. But besides this, I was aware that a memorial to the same effect as my own petition, signed by the most scientific men in London, and embracing the opinions of all the chief arctic officers, had been presented to the head of her Majesty's Government (by whom it was kindly received); and I indulged the hope that it could scarcely fail to receive your Lordships' favourable consideration.

"Thus, between doubt and hope, between occasional misgivings and reviving confidence, but withal in constant and harassing anxiety, I have passed three long months (precious months to me, who required them all for my own expedition, if that great burden were at last to fall upon me), till at last a time has arrived when the equipment of a private expedition is no longer possible, and a season of probably unexampled openness for ice navigation is passing away.

"I feel sure that if your Lordships would only do me the favour of considering for a moment the painful position in which I have thus been and am still placed, without a single word vouchsafed to me either to confirm my hopes or to extinguish them, deprived of any means but such as I had a reasonable objection to of securing public feeling in my behalf, whilst the Arctic Papers (including my appeal to your Lordships), which were called for in the House of Commons, continued to be withheld, unable thus to make use of the present or to calculate on the future, you would feel that a great hardship—nay, that a great injustice, for such I feel it to be—has been inflicted on me.

"Yet, great as this trial has been, it receives aggravation from the knowledge that I am not alone affected by it. I abstain from obtruding on you details of private matters, however they might serve to illustrate this aspect of my embarrassing position; but I feel sure that you will deem it worthy of your kind and serious attention, when I inform you that the distinguished individual who has generously offered me his gratuitous services for the command of my private expedition,* should I be unhappily reduced to this extremity, has done so at the sacrifice of all his own professional and private interests, in the purest spirit of sympathy with my anxieties and of devotion to a holy cause. And I might say much more than this, if I felt permitted to do so. Your Lordships, however, will, I am sure, perceive that I cannot indefinitely prolong the state of uncertainty in which my noble-minded and generous friend is now placed; and that it is my duty either to release him from his promise, as I would so gladly do were I sure that my cause were safe in your hands, or enable him at once to commence independent operations.

"Regretting deeply that you have, as I learn, come to a decision adverse to the immediate starting of a vessel by the eastern route, since I fully recognise the possibility of following my husband's track on that side down Peel Channel, I yet may be permitted to express the opinion I have long entertained, confirmed as it is by that of your late eminent hydrographer, Sir Francis Beaufort, and by that of Captains Collinson and Maguire, that the route by Behring Strait, though longer in distance, is of surer and safer accomplishment, and that a

* Dr E. K. Kane, of the United States Navy.

vessel despatched this autumn to Behring Strait would probably arrive at the spot to be searched in a shorter time than by the other. Captain Collinson, whose experience is the highest that can be adduced on this point, has no doubt that he could carry even such a heavy sailing-ship as the Enterprise without the aid of steam, in one season only, to the very locality where the remains of the Erebus and Terror are probably now lying, and where it is at least certain that the Esquimaux hold the secret of their fate, and of the pillage they have acquired from the catastrophe.

"This opinion of Captain Collinson as to the facility of a vessel's reaching the place of its destination in one season by way of Behring Strait, is shared by Captain Maguire, as expressed in a letter which I have permission to enclose. Your Lordships will also perceive therein another reason for the adoption of this route, which has not hitherto received the attention its extreme importance deserves, namely, the facility it gives of bringing the vessel into close contact with the Esquimaux, it being Captain Maguire's opinion (as it is that of Mr Anderson, the late commander of the boat party down the Great Fish River) that the tranquil presence of a vessel is necessary to extract the whole truth from the natives. These people are not wanting in sagacity, and if they see nothing but a boat or sledge-party, they will be sure to calculate on the very limited resources of such a party, that it will soon return whence it came, and rid them of unwelcome investigations. It is also to be recollected that the Esquimaux are in the habit of making spring and autumn migrations, so that time would be required to enable the intelligence that white men were on the coast to permeate throughout the country, and thus reach the ears of any stragglers that may yet remain of the crews of the missing vessels.

"I would entreat of your Lordships, should you doubt the accuracy of my statements, to call before you those two able and experienced officers, Captain Collinson and Captain Maguire, one of whom has brought back his ship and crew in perfect safety, after a navigation in arctic waters of unexampled length and importance; whilst the other, within a more restricted field of action at Point Barrow, succeeded so well in his endeavours to gain the confidence and co-operation of the natives, as to be an earnest of his success in any other quarter.

"I mention these two distinguished officers as being especially qualified to speak of the advantages and disadvantages of the route suggested, not forgetting that Captains Osborn and Richards are also on the spot, equally able to submit to your Lordships, if honoured by your reference, all that might have been said, in favour or otherwise, of the route which you have pronounced to be impracticable at this advanced season. All are alike ignorant that I am expressing this unbounded confidence in their capacity and zeal, in the humble hope of reminding your Lordships that if you give little weight to anything I can advance, as coming from an incompetent or too interested person, there are those at hand whose qualifications, whose duty towards you, and whose sense of responsibility, remove them widely from such disparaging circumstances.

"Whilst this subject is still under deliberation, I commit the prayer of my present appeal to your serious and humane consideration, believing that the honour of my country is no less concerned in the result than are my own personal interests and those of my fellow-sufferers in calamity.

"I have the honour to be, my Lords, your obedient servant,

(Signed) "JANE FRANKLIN.

"To the Lords Commissioners
of the Admiralty."

MEMORIAL

PRESENTED BY

SIR RODERICK IMPEY MURCHISON,

G.C.S.S.; D.C.L.; M.A.; F.R.S.; F.L.S.; HON. MEM. R.S. ED.; R.I.A.

"*To the Right Hon.* VISCOUNT PALMERSTON, *M.P.*, *G.C.B.*

"LONDON, *June* 5.

"Impressed with the belief that her Majesty's missing ships, the Erebus and Terror, or their remains, are still frozen up at no great distance from the spot whence certain relics of Sir John Franklin and his crews were obtained by Dr Rae—we whose names are undersigned, whether men of science and others who have taken a deep interest in arctic discovery, or explorers who have been employed in the search for our lost countrymen, beg earnestly to impress upon your Lordship the desirableness of sending out an expedition to satisfy the honour of our country, and clear up a mystery which has excited the sympathy of the civilised world.

"This request is supported by many persons well versed in arctic surveys, who, seeing that the proposed expedition is to be directed *to one limited area only*, are of opinion that the object is attainable, and with little risk.

"We can scarcely believe that the British Government, which, to its great credit, has made so many efforts in various directions to discover even the route pursued by Franklin, should cease to prosecute research, now that the locality has been clearly indicated where the vessels or their remains must lie—including, as we hope, records which will throw fresh light on arctic geography, and dispel the obscurity in which the voyage and fate of our countrymen are still involved.

"Although most persons have arrived at the conclusion that there can now be no survivors of Franklin's expedition, yet there are eminent men in our own country, and in America, who hold a contrary opinion. Dr Kane of the United States, for example, who has distinguished himself by pushing farther to the north in search of Franklin than any other individual, and to whom the Royal Geographical Society has recently awarded its founders' gold medal, thus speaks (in a letter to the benevolent Mr Grinnell) :—'I am really in doubt as to the preservation of human life. I well know how glad I would have been, had my duty to others permitted me, to have taken refuge among the Esquimaux of Smith Strait and Etah Bay. Strange as it may seem to you, we regarded the coarse life of these people with eyes of envy, and did not doubt but that we could have lived in comfort upon their resources. It required all my powers, moral and physical, to prevent my men from deserting to the walrus settlements, and it was my final intention to have taken to Esquimaux life had

Providence not carried us through in our hazardous escape.'

"But passing from speculation, and confining ourselves alone to the question of finding the missing ships or their records, we would observe that no land expedition down the Back River, like that which, with great difficulty, recently reached Montreal Island, can satisfactorily accomplish the end we have in view. The frail birch-bark canoes in which Mr Anderson conducted his search with so much ability, the dangers of the river, the sterile nature of the tract near its embouchure, and the necessary failure of provisions, prevented the commencement, even, of such a search as can alone be satisfactorily and thoroughly accomplished by the crew of a man-of-war—to say nothing of the moral influence of a strongly armed party remaining in the vicinity of the spot until the confidence of the natives be obtained.

"Many arctic explorers, independently of those whose names are appended, and who are absent on service, have expressed their belief that there are several routes by which a *screw*-vessel could so closely approach the area in question as to clear up all doubt.

"In respect to one of these courses, or that by Behring Strait, along the coast of North America, we know that a single sailing-vessel passed to Cambridge Bay, within 150 miles of the mouth of the Back River, and returned home unscathed, its commander having expressed his conviction that the passage in question is so constantly open that ships can navigate it without difficulty in one season. Other routes, whether by Regent Inlet, Peel Sound, or across from Repulse Bay, are preferred by officers whose experience in arctic matters entitles them to every consideration; whilst in reference to two of these routes it is right to state that vast quantities of provisions have been left in their vicinity.

"Without venturing to suggest which of these plans should be adopted, we earnestly beg your Lordship to sanction without delay such an expedition as, in the judgment of a committee of arctic voyagers and geographers, may be considered best adapted to secure the object.

"We would ask your Lordship to reflect upon the great difference between a clearly-defined voyage to a narrow and circumscribed area, within which the missing vessels or their remains must lie, and those former necessarily tentative explorations in various directions, the frequent allusions to the difficulty of which, in regions far to the north of the voyage now contemplated, have led persons unacquainted with geography to suppose that such a modified and limited attempt as that which we propose involves farther risk, and may call for future researches. The very nature of the former expeditions exposed them, it is true, to risk, since regions had to be traversed which were totally unknown; while the search we ask for is to be directed to a circumscribed area, the confines of which have already been reached without difficulty by one of her Majesty's vessels.

"Now, inasmuch as France, after repeated fruitless efforts to ascertain the fate of La Pérouse, no sooner heard of the discovery of some relics of that eminent navigator, than she sent out a searching expedition to collect every fragment pertaining to his vessels, so we trust that those arctic researches which have reflected much honour upon our country may not be abandoned at the very moment when an explanation of the wanderings and fate of our lost navigators seems to be within our grasp.

"In conclusion, we further earnestly pray that it may not be left to the efforts of individuals of another and kindred nation already so distinguished in this cause, nor yet to the noble-minded widow of our lamented friend, to make an endeavour which can be so much more effectively carried out by the British Government.

"We have the honour to be, &c.,

"F. Beaufort, R. I. Murchison, F. W. Beechey, Wrottesley, E. Sabine, Egerton Ellesmere, W. Whewell, R. Collinson, W. H. Sykes, C. Daubeny, J. Fergus, P. E. de Strzelecki, W. H. Smyth, A. Majendie, R. Fitz Roy, E. Gardiner, Fishbourne, R. Brown, G. Macartney, L. Horner, W. H. Fitton, Lyon Playfair, T. Thorp, C. Wheatstone, W. J. Hooker, J. D. Hooker, J. Arrowsmith, P. La Trobe, W. A. B. Hamilton, R. Stephenson, J. E. Portlock, C. Piazzi Smyth, C. W. Pasley, G. Rennie, J. P. Gassiot, G. B. Airy, J. F. Burgoyne."

TABLE showing the MEAN HEIGHT of BAROMETER, with the Temperature of the Air, on board H.M.S. Investigator, from August 1850 to March 1853.

	Year and Month.	Barometer. Maximum.	Barometer. Minimum.	Barometer. Mean.	Temperature of Air. Maximum.	Temperature of Air. Minimum.	Temperature of Air. Mean.	Mean Force of Wind.	Yearly Abstract.
	1850.								
PRINCE OF WALES STRAIT.	Aug.	30·060	29·390	29·751	+50	+27	+36·5	3·5	Barometer. Max. 30·650; Min. 29·160; Mean, 29·828. Air. Max. +5; Min. —40; Mean, —4·66.
	Sept.	·650	·470	·809	+46	—1	+20·2	3·6	
	Oct.	·180	·380	·861	+24	—23	+0·2	2·0	
	Nov.	·270	·160	·739	+7	—32	—10·2	3·1	
	Dec.	·560	·480	·978	—4	—40	—23·4	2·5	
	1851.								
	Jan.	·570	·400	·885	—15	—51	—32·5		Barometer. Maximum . 30·750 Minimum . 29·030 Mean . . . 29·934 Air. Maximum . +52·0 Minimum . —51·0 Mean . . . +2·58
	Feb.	·630	·030	·958	—9	—51	—37·7		
	Mar.	·720	·338	·946	—5	—51	—28·8		
	April.	·610	·410	30·037	+38	—32	—4·8	3·1	
	May.	·600	·560	·023	+47	—5	+18·9	2·2	
In MERCY BAY, BANKS LAND, lat. 70° 6′ N. long. 118° 15′ W.	June	·150	·470	29·837	+53	+27	+36·1	3·5	
	July.	·090	·450	·756	+52	+32	+37·5	3·0	
	Aug.	·400	·390	·865	+52	+21	+37·6	2·8	
	Sept.	·270	·450	·876	+43	+1	+24·6	3·1	
	Oct.	·200	·300	·877	+26	—22	+3.3	1·9	
	Nov.	·750	·630	30·097	+10	—40	—15·2	1·8	
	Dec.	·810	·490	·046	+11	—44	—20·0	3·5	
	1852.								
	Jan.	·600	·280	29·841	+8	—51	—27·3	3·4	Barometer. Maximum . 31·000 Minimum . 28·970 Mean . . . 29·906 Air. Maximum . +52 Minimum . —52 Mean . . . +0·05
	Feb.	31·000	·070	·777	—1	—47	—25·8	3·1	
	Mar.	·000	·410	30·082	+5	—52	—28·4	2·0	
	April.	30·430	·520	·164	+31	—38	—1·4	2·5	
	May.	·250	·600	29·987	+37	—25	+10·2	2·6	
	June.	·100	·430	·758	+51	+11	+31·5	3·1	
	July.	·000	·370	·749	+52	+30	+36·7	2·9	
	Aug.	·170	·400	·816	+52	+19	+33·2	2·9	
	Sept.	·100	·070	·785	+38	—4	+20·1	3·6	
	Oct.	·300	·440	·986	+16	—33	—5·6	2·2	
	Nov.	·680	·460	·978	+9	—43	—16·5	3·1	
	Dec.	·670	28·970	·944	—4	—48	—26·1	3·7	
	1853.								
	Jan.	·120	29·180	·748	—16	—65	—43·87	4·05	Barometer. Max. 30·72; Min. 29·180; Mean, 29·960. Air. Max. +17; Min. —65; Mean, —35·92.
	Feb.	·580	·400	30·085	—13	—57	—38·50	2·50	
	Mar.	·720	·540	·048	+17	—58	—25·4	2·30	

ROBERT M'CLURE, Commander,
Her Majesty's Ship Investigator.

GAME killed in the ARCTIC REGIONS by the Crew of H.M.S. Investigator.

	Number killed.	Average Weight each.	Total Weight.
Musk-oxen, . .	7	278 lb.	1945 lb.
Deer, ...	110	70 ,,	7716 ,,
Hares, ...	169	6 ,,	1014 ,,
Grouse, ...	186	Not weighed.	..
Ducks, ...	198	,,	..
Geese, ...	29	,,	..
Wolves, . . .	2	,,	..
Bears, ...	4	,,	..

Total head killed, ... 705

LIST of GAME procured by H.M.S. Resolute and Intrepid in the ARCTIC REGIONS, between September 1852 and September 1853, by Captain F. L. M'Clintock, R.N.

Locality.	Musk-Oxen.	Reindeer.	Hares.	Bears.	Wolves.	Foxes.	Seals.	Ptarmigan.	Water-Fowl.
Almost exclusively on Melville Island.	112	94	161	4	3	52	2	684	388
	Average Weight when cleaned for the Table—								
	116 lb.	60 lb.	7 lb.					1 lb.	2½ lb.

Total Weight of Meat procured, ... 28,284 lb.

APPENDIX.

ON THE OCCURRENCE OF NUMEROUS FRAGMENTS OF FIR-WOOD IN THE ISLANDS OF THE ARCTIC ARCHIPELAGO; WITH REMARKS ON THE ROCK-SPECIMENS BROUGHT FROM THAT REGION. BY SIR RODERICK IMPEY MURCHISON, D.C.L., F.R.S., V.P.G.S., DIRECTOR-GENERAL OF THE GEOLOGICAL SURVEY.

I CANNOT attempt to offer any general, still less any detailed description of the rocks and fossils of the north-western portion of that great arctic archipelago whose shores were first explored by Parry and Sabine. The specimens they brought home from Melville Island, and which were described by Mr Kōnig, first conveyed to us the general knowledge of the existence there of fossiliferous limestones and other rocks analogous to known European types in Scandinavia. Since those early days, the voyages of Franklin, and of the various gallant officers who have been in search of our lamented friend, have amplified those views, and have shown us that over nearly the whole of the arctic archipelago these vast islands possess a structure similar to that of North America. My chief object now is to call attention to the remarkable fact of the occurrence of considerable quantities of wood, capable of being used for fuel or other purposes, which exist in the interior, and on the high grounds of large islands in latitudes where the dwarf willow is now the only living shrub.

Before I allude to this phenomenon, as brought to my notice by Captain M'Clure and Lieutenant Pim, I would, however, briefly advert to a few rock-specimens collected by the latter officer and his comrades in Beechey Island, Bathurst Land, Eglintoun Island, Melville Island, Prince Patrick Island, and Banks Land, where he joined Captain M'Clure—specimens which we ought to value highly, seeing that they were saved from loss under very trying circumstances.

From this collection, as well as from other sources to which I have had access, as derived from the voyages of Parry, Franklin, Back, Penny, Inglefield, and the recent work of Dr P. Sutherland, I am led to believe that the oldest fossiliferous rock of the arctic region is the Upper Silurian—viz., a limestone identical in composition and organic contents with the well-known rocks of Wenlock, Dudley, and Gothland

No clear evidence has been afforded as to the existence of Devonian rocks, though we have heard of red and brownish sandstone, as observed in very many localities by various explorers, and which possibly may belong to that formation. Thus, in North Somerset, to the south of Barrow Strait, red sandstone is associated with the older limestones. Byam Martin Island was described by Parry as essentially composed of sandstone, with some granitic and felspathic rocks; and whilst the north-eastern face of Banks Land is sandstone, its north-western cliffs consist (as made known by Captain M'Clure) of limestone. But whilst in the fossils we have keys to the age of the Silurian rocks, we have as yet no adequate grounds whereon to form a rational conjecture as to the presence of the Old Red Sandstone, or Devonian group.

True Carboniferous Producti and *Spiriferi* have been brought home by Sir E. Belcher from Albert Land, north of Wellington Channel, and fossil plants have been collected by Captain Sherard Osborn; and hence we may affirm positively, that the old Carboniferous rocks are also present. Here and there bituminous schist and coal are met with; the existence of the latter being marked at several points on the general chart published by the Admiralty. With the palæozoic rocks are associated others of igneous origin and of crystalline and metamorphosed character. Thus, from Eglintoun Island to the south of Prince Patrick Island, first defined by the survey of Captain Kellett and his officers, we see concretions of greenstone, associated with siliceous or quartzose rocks and coarse ferruginous grits; and in Princess-Royal Island, besides the characteristic Silurian limestones, there are black basalts and red jaspers, as well as red rocks, less altered by heat, but showing a passage into jasper. Highly crystalline gypsum was also procured by Lieutenant Pim from the northern shores of Melville Island. In the collection before us there are silicified stems of plants, which Lieutenant Pim gathered on various points between Wellington Channel on the east and Banks Land on the west. Similar silicified plants were also brought home by Captain M'Clure from Banks Land; and through the kindness of Mr Barrow, to whom they were presented, they are now exhibited, together with a collection made by Captain Kellett, which he sent to Dr J. E. Gray of the British Museum, who has obligingly lent them for comparison.

I had requested Dr Hooker to examine all those specimens which had passed through my hands, and I learn from him that he will prepare a description of them, as well as of a great number from the same region, which had been sent to his father, Sir W. Hooker, associated, like those now under consideration, with fragments of recent wood.

Of Secondary formations no other evidence has been met with, except some fossil bones of Saurians, brought home by Sir E. Belcher, from the smaller islands north of Wellington Channel. Of the old Tertiary rocks, as characterised by their organic remains, no distinct traces have, as far as I am aware, been discovered; and hence we may infer that the ancient submarine sediments, having been elevated, remained during a very long period beyond the influence of depositary action.

Let us now see how the other facts, brought to our notice

by the gallant arctic explorers who have recently returned to our country, bear upon the relations of land and water in this arctic region during the quasi-modern period, when the present species of trees were in existence.

Captain M'Clure states that in Banks Land, in latitude 74° 48′, and thence extending along a range of hills varying from 350 to 500 feet above the sea, and from half a mile to upwards inland, he found great quantities of wood, some of which was rotten and decomposed, but much of it sufficiently fresh to be cut up and used as fuel. Whenever this wood was in a well-preserved state, it was either detected in gullies or ravines, or had probably been recently exhumed from the frozen soil or ice. In such cases, and particularly on the northern faces of the slopes where the sun never acts, wood might be preserved any length of time, inasmuch as Captain M'Clure tells me he has eaten beef, which, though hung up in his cold larder for two years, was perfectly untainted.

The most remarkable of these specimens of well-preserved recent wood is the segment of a tree, which, by Captain M'Clure's orders, was sawn from a trunk sticking out of a ravine, and which is now exhibited.* It measures 3 feet 6 inches in circumference. Still more interesting is the cone of one of these fir-trees which he brought home, and which apparently belongs to an *Abies* resembling *A. alba*, a plant still living within the arctic circle. One of Lieutenant Pim's specimens of wood from Prince Patrick Island is of the same character as that just mentioned, and in its microscopical characters much resembles *Pinus strobus*, the American pine, according to Professor Quekett, who refers another specimen, brought from Hecla and Griper Bay, to the larch.

In like manner similar fragments of wood were seen two degrees further to the north, in Prince Patrick Land, and also

* Through the kindness of Mr John Barrow, to whom it had been given, this wood, with some silicified stems, has been presented to the Museum of Practical Geology.

in ravines of the interior of that island, where, as I am informed, a fragment was found, like the tree described by M'Clure, protruding from the soil on the side of a gully.

We learn, indeed, from Parry's 'Voyage,' that portions of a large fir-tree were found at some distance from the south shore of Melville Island, at about 30 feet above high-water mark, in latitude 74° 59′ and longitude 106°.* According to the testimony of Captain M'Clure and Lieutenant Pim, all the timber they saw resembled the present driftwood so well known to arctic explorers, being irregularly distributed, and in a fragmentary condition, as if it had been broken up and floated to its present positions by water.

If such were the method by which the timber was distributed, geologists can readily account for its present position in the interior of the arctic islands. They infer that at the period of such distribution large portions of these tracts were beneath the waters, and that the trees and cones were drifted from the nearest lands on which they grew. A subsequent elevation, by which these islands assumed their present configuration, would really be in perfect harmony with those great changes of relative level which we know to have occurred in the British Isles, Germany, Scandinavia, and Russia since the glacial period. The transportation of immense quantities of timber towards the north pole, and its deposit on submarine rocks, is by no means so remarkable a phenomenon as the wide distribution of erratic blocks during the glacial epoch over northern Germany, central Russia, and large portions of our island when under water, followed by the rise of these vast masses into land. If we adopt this explanation, and look to the extreme cold of the arctic region in the

* "Sergeant Martin of the Artillery, and Captain Sabine's servant, brought down to the beach several pieces of a large fir-tree, which they found nearly buried in the sand at the distance of 330 or 400 yards from the present high-water mark, and not less than 30 feet above the level of the sea."—*Parry's Voyage for the Discovery of the North-West Passage*, p. 68.

comparatively modern period during which this wood has been drifted or preserved, we can have no difficulty in accounting for the different states in which timber is found. Those portions of it which happen to have been exposed to the alternations of frost and thaw, and the influence of the sun, have necessarily become rotten; whilst all those fragments which remained enclosed in frozen mud or ice which have never been melted, would, when brought to light by the opening of ravines or other accidental causes, present just as fresh an appearance as the specimens now exhibited.

The only circumstance within my knowledge which militates against this view is one communicated to me by Captain Sir Edward Belcher, who, in lat. 75° 30′, long. 92° 15′, observed on the east side of Wellington Channel the trunk of a fir-tree standing vertically, and which, being cleared of the surrounding earth, &c., was found to extend its roots into what he supposed to be the soil.

If from this observation we should be led to imagine that all the innumerable fragments of timber found in these polar latitudes belonged to trees that grew upon the spot, and on the ground over which they are now distributed, we should be driven to adopt the anomalous hypothesis, that, notwithstanding physical relations of land and water similar to those which now prevail (*i.e.*, of great masses of land high above the sea), trees of large size grew on such *terra firma* within a few degrees of the north pole!—a supposition which I consider to be wholly incompatible with the data in our possession, and at variance with the laws of isothermal lines.

If, however, we adopt the theory of a former submarine drift,* followed by a subsequent elevation of the sea-bottom,

* Dr Hooker informs me that all the specimens sent to him were collected in mounds of silt, rising up from the level of the sea to 100 feet or more above it; and he entirely coincides with me in the belief that the whole of this timber was drifted to the spots where it now lies.

as easily accounting for all the phenomena, we may explain the curious case brought to our notice by Sir Edward Belcher, by supposing that the tree he uncovered had been floated away with its roots downwards, accompanied by attached and entangled mud and stones, and lodged in a bay, like certain "snags" of the great American rivers. Under this view, the case referred to must be considered as a mere exception, whilst the general inference we naturally draw is, that the vast quantities of broken recent timber, as observed by numerous explorers, were drifted to their present position when the islands of the arctic archipelago were submerged. This inference is indeed supported by the unanswerable evidence of the submarine associates of the timber; for, from the summit of Coxcomb Range in Banks Land, and at a height of 500 feet above the sea, Captain M'Clure brought home a fine large specimen of *Cyprina Islandica*, which is undistinguishable from the species so common in the glacial drift of the Clyde;* whilst Captain Sir E. Belcher found the remains of *whales* on lands of considerable altitude in lat. 78° north.

Reasoning from such facts, all geologists are agreed in considering the shingle, mud, gravel, and beaches in which animals of the arctic region are imbedded in many parts of northern Europe, as decisive proofs of a period when a glacial sea covered large portions of such lands; and the only distinction between such deposits in Britain and those which were formed in the arctic circle, is that the wood which was transported to the latter has been preserved in its ligneous state for thousands of years, through the excessive cold of the region.

* In Parry's 'Voyage' (p. 61) we learn that a number of marine shells, of the Venus tribe, were found imbedded in the ravines of Byam Martin Island; a fact which strengthens the view here adopted of the submergence of large portions of these tracts at a very recent geological epoch.

NARRATIVE OF COMMANDER MAGUIRE, WINTERING AT POINT BARROW.

In accordance with my last communication, I proceeded to sea from Port Clarence on the morning of the 21st of August, and with a favourable breeze passed through Behring Strait by the eastern passage on the following day at noon. A succession of contrary winds delayed our progress to the N.E. to a much greater extent than was considered favourable to insure our complete success of rounding Point Barrow at that advanced period of the season.

In our passage to the northward we passed several whale-ships cruising in squadrons, a caution they seem to have prudently adopted, for the benefit of affording mutual assistance in the event of disaster. Their success up to that time seemed to be indifferent; and we have been since informed by natives from Point Hope that whales have become very scarce on the coast since the ships have come in pursuit of them. The last whale-ship (French) seen by us, was on the morning of the 25th of August, in lat. 69° 30′ N., long. 167° 43′ W., carrying all sail to the southward. We soon afterwards made the ice in heavy floes, and tacked inshore to ascertain its distance from the land, when we found the contrary winds had done us good service by opening a free passage of from ten to fifteen miles, in which we beat to the N.E., making but slow progress until the night of the 2d of September, when a slant from the southward, with a fast-falling barometer, warned us that a change of weather was at hand. Our distance from Point Barrow, now reduced to fifty miles, I thought we could accomplish before the ice set inshore, and therefore pushed forward under all sail and rounded it at the distance of one mile, on the following day at noon, September 3d.

The approach to the channel leading between the sandy islets that form the protection we were about to seek for the winter in Elson Bay, was found, contrary to our expectations, shoal and intricate, making it necessary to anchor the ship and sound out the passage. If it was found not to afford sufficient water, of which there was a doubt, our position was not one to lose time in, shut out, and close down on an exposed shore, with a gale coming on that would soon have loaded it with ice. Having ascertained, as expeditiously as possible, that there were about nine inches to spare across a shoal before we got to deep water, the anchor was weighed; and after making a few tacks, the narrowness of the channel, and the ship taking the ground twice, made it advisable to anchor and kedge under shelter of the spit. A fortunate turn in the current enabled us to effect this, as by the time the warps had been run out, the gale had increased so much as to render it unsafe to trip the anchor; however, finding a strong weather-current setting, it was weighed, and the ship warped into a wild-looking anchorage for protection, in a gale of wind, no land being visible, except the low sand-spit of Point Barrow and the islet adjoining, not more than five feet above the level of the sea, which broke over them with great violence during the height of the gale. These are again guarded by the shoals lying off, on which the drift-ice grounds, making the anchorage, when gained, secure, but difficult of access or egress. Daylight next morning showed us how fortunate we had been in getting shelter. The gale, now veered to west, was unabated; the sea broke heavily over the shoals passed yesterday, and against the sand-spits to within a short distance of the ship, whilst the offing was encumbered with heavy ice, becoming gradually closer with the gale. As we found a strong current setting to windward, I had no doubt of the ship holding on, although the confined space of the anchorage did not admit veering more than thirty fathoms of cable. In the afternoon the gale began to moderate, and on the following day the ship was moved to a more secure berth, near the position selected for winter-quarters, where we remained until the ice set fast on the 24th of September.

A succession of strong gales and thick weather, for the following week, retarded our preparations, consisting chiefly in collecting driftwood, not found here in any abundance. From this material the plank for housing-in was sawed, and the remains stacked for the winter firing. Advantage was also taken of every opportunity to send a boat to sound the channel, knowing the difficulty of the task after the ice had formed; and on its breaking up we should be too anxious to move with it to have time for that purpose. On the 25th pancake ice began to form in the bay, and drift out rapidly with the current. A party was sent to haul the launch up on the adjacent islet, to be out of reach of the natives; this service was performed by Mr G. T. Gordon, mate, who, when returning in the gig with a fresh and favourable wind, was unable to push his way through the young ice, and was carried in it through the passage into the offing. In this distressing dilemma a second boat was lowered, in which Lieutenant Vernon promptly volunteered his services; and by running out 700 fathoms of whale-line the gig was reached, now carried some distance off the land. By this means they were enabled to reach the spit, although they had another narrow escape from being carried out, by the line parting when they were close to it; fortunately, one of the men was sufficiently quick to heave the end amongst a crowd of Esquimaux, drawn to the spot by witnessing the state of our boats; and they hauled them up, where they had to remain for the night, the ice being now too strong to allow us to haul the boats through it, and not sufficiently firm for the people to walk on board. During the night, the ice moved but once a short distance, then set fast; and in the morning we had the satisfaction of receiving our boat's crew on board, after experiencing a degree of anxiety for their safety that is not easily described.

The following days were occupied in sawing a canal towards

our winter position, which was much delayed by unexpected movements in the ice, undoing our work when nearly completed; and on the 30th, at 10 P.M., we were tracked up it by about seventy natives, men, women, and children, whose shouts and exclamations of surprise gave animation to the whole scene, and made it one of deep interest.

Our time was now busily occupied in making the usual preparations for passing a winter in this rigorous climate, which we had thus early observed symptoms of. A temporary house was erected close to the ship to receive our deck-load of provisions, to enable us to have them clear for the crew to take exercise when the state of the weather would not admit of their leaving the ship; and an observatory, for the reception of the magnetical instruments, was constructed from ice alone, which answered the purpose perfectly for eight months.

These arrangements were completed by the 20th of October, when the necessary winter routine was established for an economical expenditure of fuel and provisions, with due attention to order, cleanliness, occupation, and amusement, to lighten as much as possible a time confessed by all as being depressing and monotonous. Many valuable hints on this subject were gained from the works of Captain Parry, in following whose example I consider we could not err. Taking advantage of his experience, the masts and yards were kept in their proper places, affording a better mark for seeking the ship from a distance, bearing in mind our being here in expectation of parties falling back upon us for safety; and as the land is very low, and in winter, it may be said, not visible, the ship made a fine object, being discernible in clear weather at the distance of nine miles from every direction.

Deeming it a matter of importance that the Plover's position at Point Barrow should be known as far to the eastward as possible, and also wishing to ascertain whether Dease Inlet would afford shelter for any vessel that might at any future time be desirous of wintering there, I took the earliest opportunity of making a boat excursion to perform this service, and left the ship on the morning of the 21st of September, in the gig, accompanied by Mr T. A. Hull, second master. Steering off the land into five fathoms, we passed a good deal of sailing ice, and proceeded E.N.E. over an even bottom of from five to six fathoms with the wind from the S.E.

Changing our course to the southward, we got entangled among a series of sand-spits, when, taking to our oars, we steered along the outside of the largest island of the group, which we then supposed to be connected with the mainland about Point Christie; but it was afterwards found to be one of the very low chain of sandy islets running along this coast. As I suspected this was taking us to the eastward of the inlet, I landed to ascend the highest part, to see how much further it extended, and found a second island running in the same direction. The water being too shoal for hauling the boat up here, we stood out into deeper water, and at length succeeded in hauling the boat up for the night on the second island. During the night a strong breeze sprang up from the N.W., with a temperature of +30°.

Knowing we had run our distance for Dease Inlet, I was not a little puzzled to know where we were, as I could scarcely fancy it was possible to lay down this coast without noticing these islands.

On the following morning, our observations being complete and the boat loaded, and now concluding that Dease Inlet must be looked for to the southward, and first erecting a conspicuous mark on this place of the Plover's winter position, we steered for a point of the main just visible to S.W. (true).

This proved to be Point Christie, where we landed in time to get the latitude at noon. Its higher part does not exceed ten feet above the level of the sea; and here we erected another large mark. The season now seemed to be so far advanced that I was in doubt between crossing the inlet and returning to the ship; but, as I considered my object would not be carried out without placing notices on Point Tangent, I determined to cross it, and started with a fine leading wind from the northward. In two hours we reached the eastern shore, which is even more shoal than the western—the water about Point Tangent being so shallow that our boat could not be got within a cable's length of the beach.

Having now found that the greatest depth of water to be obtained by sounding directly across Dease Inlet was eleven feet only, with its shores extremely shoal, I considered the question settled that no vessel could find winter-quarters there; and, after leaving the proper notices, started on our return to the ship. Sludge ice was observed on our return to be forming in all the small bays, the temperature having fallen to +19°, warning us that the open season was nearly at an end. Passing the night in the same place as the previous one, for the sake of the driftwood, we left the next morning with all haste for the ship, which, with a fine breeze from the northward, we reached by noon. Twelve hours after our return, it was reported to me that the ice was drifting past the ship.

A further examination of Dease Inlet was afterwards made by Mr T. A. Hull, second master, in the month of May, in continuation of a survey of this coast from Point Barrow eastward.

The southern shores of this inlet, which had hitherto been left blank on the chart, were now traced. It was found to extend in a S.W. direction for a distance of twenty miles, its breadth at the mouth being eight miles, and terminating in a shoal bay. The S.E. shore is much higher than the rest, one cliff there being as high as twenty-four feet. Four inconsiderable rivers empty themselves here—two on the eastern, and two on the western shore.

The chain of islands which, commencing at the Plover's winter-quarters, closely abut on Point Tangent, the western part of which was discovered by Captain Moore, and denominated by him Plover Group, has been found to be ten in number (on but two or three of the largest of which is there the slightest sign of vegetation), running in a line almost parallel with that of the coast, or E.S.E. and W.N.W. (true) from Point Barrow to Point Tangent, where they terminate, the only channel between them, of sufficient depth for a ship, being the one by which the Plover entered.

From the time of our arrival at winter-quarters, situated two miles E.S.E. (true) from the Esquimaux settlements on Point Barrow, called by them Noowook, we found this people, contrary to our preconceived opinion, very troublesome and unfriendly. To such an extent did this feeling exhibit itself, that it would have been prudent to remove from their vicinity (particularly as we had received more than one unmistakable hint to that effect), had circumstances admitted it; but, as we occupied the only spot of deep water to be found on this part of the coast, it became necessary to put up with the evil, hoping that time and a better knowledge of our character would improve their conduct; and I had no doubt our wintering amongst them would eventually be attended with beneficial results. The commencement of our intercourse was attended with many unpleasant circumstances. No single boat's crew could be at any distance from the ship without being pilfered from in the most daring and barefaced way; and upon every trivial, and often without any, occasion, their knives were drawn upon our men, who, although armed with muskets, had strict orders in no case to make even a show of them, unless obliged by necessity, as I thought recourse to that force was to be avoided when a good feeling in favour of any of our missing countrymen, who may at any future period be in their power, was the object sought. Carrying out these views to the extent of not showing our arms was not appreciated, as they mistook forbearance for timidity; and, at the request of two officers going with a watering party to the village, to carry their guns nominally for the purpose of shooting small birds, the show of them was found to have so good an effect that it was adopted on all future occasions, although we were obliged to cease sending for water after a few turns, there being always some unpleasant display of feeling on their part that was best avoided when possible.

Whilst occurrences such as I have mentioned were taking place daily with our parties away from the ship, the difficulty of dealing with those collected about her was sufficient to employ all the people left on board, exclusive of the pressing duties of the ship, with a small crew, at this season.

About the 15th of September, they appeared to be returning to their winter huts, from their usual summer's excursion along the coast to the eastward, and, as the ship lay in their direct track, we had a visit from all of them, including also the Cape Smyth tribe, being the two most numerous on any part of the coast, numbering together about five hundred. As many as seven or eight large u-mi-aks arrived daily for eight or nine days, containing their summer tents, families, dogs, and sledges, &c.: they appeared perfect strangers, and looked in amazement at us and the ship. They brought with them a small quantity of fish and venison: with the latter they parted reluctantly, and seemed to prefer begging and stealing (in which they were most unscrupulous) to any kind of exchange.

On the morning of the 17th I was informed that a large u-mi-ak had come alongside, and the crew had forced their way on board. As this was not an uncommon case, I thought nothing of it when I found that Lieutenant Vernon was attending on deck. He soon came down to inform me that the chief of the party had a musket, and was very anxious to get gunpowder in exchange for venison. This piece of information I considered the worst I had received, amongst many unpleasant circumstances that I had experienced, feeling that we could not remain amongst them if they had fire-arms.

It will be proper to state here that we have, at a very late period of our stay, identified this chief as the same who followed and annoyed Commander Pullen at Point Berens in 1849, full particulars of which are given in his journal.

As he expressed a wish to see me, I went up, and found a large, powerful, elderly man, with a peculiarly bad expression of countenance. He had a Hudson Bay musket, with the name of Barnett on the lock; it was a good deal worn, but fit for service. He had a powder-horn hanging hunter-fashion under his left arm, but pretended to have neither ball nor shot, for which he was most pressing, and would not dispose of anything except for ammunition. This, as a matter of course, he was not supplied with. I saluted him with much friendship, made his wife a present, and took him down to my cabin, where I made him a present of tobacco, and satisfied his curiosity about the ship below. Then I took him on deck, with the idea that he would go away; but nothing seemed further from his thoughts, as he remained about the decks and slipped down the hatchways on to the lower deck several times—a part of the ship they had not had access to during any period of our stay. During the forenoon several u-mi-aks arrived alongside the ship, discharging their crews in swarms on our deck, so as literally to crowd it for the day. They were allowed every freedom consistent with their known propensity for stealing; but some, bolder than others, were difficult to deal with. One man attempted to force back the after ladder-doors, and my stopping him brought about a slight scuffle between us. That did not seem to have satisfied him, as he soon afterwards came in contact with the quartermaster of the watch, a quiet but rather short-tempered, powerful young man, who, before anybody could interfere, gave him a lesson he will not soon forget; he dealt him fair English blows about the head, each of them sufficient to stun any one except an Esquimaux; but he received them until they had the effect of quite taming him, when he was put over the side in the presence of at least sixty of his countrymen, few of whom offered to interfere, and the remainder looked on with indifference. About noon, when, at my particular desire, three parts of the crowd went away, the remainder were evidently detained by the old chief, whom there was no moving out of the ship without having recourse to force; and this I had no intention of, preferring to wait until he got tired of his visit, and this seemed unlikely for the present, as he hailed three u-mi-aks full of people to come alongside. I heard the word "tawac" (tobacco) used very often, I suppose as the inducement, and the children I observed had been sent away. It occurred to me they might have thoughts of pillaging the ship; their numbers to ours seeming so overpowering. In order to be prepared for anything of the kind, the men stationed on deck were sent down, one at a time, to arm themselves with pistols, to be kept out of sight in their breasts, in the event of a simultaneous attack being made with their knives, all being provided with good ones, and adepts in their use. When the men were all armed, I was satisfied to await the result. A silence seemed to prevail, as if they had not decided what to do; and whether they had or had not meditated any mischief, beyond stealing as much as they could, they attempted nothing, and went away as night came on, leaving the old chief with his own boat only. He had continued to range about the ship in the most insolent way: and I think it reasonable to suppose it was only the fear of our fire-arms that kept him from mischief. When left by himself, I was cautious not to urge his going away, as I had done when there were seventy people with him; but letting him choose his own time, he remained until 7 P.M.—a visit of twelve hours. When he was gone, I was so thoroughly tired and provoked, and knowing that every person in the ship must be suffering in the same way, that it became necessary to adopt a different system, the number of small articles stolen during the day, notwithstanding all our vigilance, affording sufficient pretext for the change. All work was stopped the next day, and an efficient arrangement made to allow only one boat's crew on board at a time; and whatever dissatisfaction it might give, it was necessary to adopt it.

It seems necessary to mention here the difficulty of keeping a numerous tribe of natives out of a vessel like the Plover, as the ice-chocks made a convenient landing-place on the outside, not more than four feet from the water, running the whole length on both sides, where they mounted in all directions, and in some cases, when prevented, they cut at our men's legs with their knives, and in one or two instances cut through box-cloth trousers. Whilst the knives of some were engaged in this way, those of others were busily employed cutting the lead scupper-pipes out of the side; the nails of the copper were proof against them, but no part of the side escaped their attempts. From this cause it became necessary to cover all the parts assailable outside with a sheathing of wood; and after the ship was frozen in she was enclosed round with a chain, rove through posts fixed in the ice at the distance of seven yards from the side. This arrangement, although very unpopular, was found most beneficial.

The day succeeding the one last described, the chief was observed sitting on the spit close to the ship, and, I was told, had hailed the ship as if he wished to be sent for. This I thought too good a joke. He was soon afterwards picked up by his own boat and came alongside, but, to his surprise, was not allowed on board, as so many things had been stolen the previous day. During the time he was standing on the gangway, the crew happened to be cleaning, discharging, and reloading their arms, and examining the two carronades, at which he seemed to stare a good deal and went away. Several u-mi-aks arrived alongside, as usual, during the day; but none of the crews were allowed on board. Some bartering was carried on, and a few presents made to them; but they seemed to prefer adding to their stock by stealing, to any exchange.

The following morning we had another early visit from the chief, accompanied by some other leading men. They brought, as a peace-offering, all the articles stolen from the ship for several days. This I considered very satisfactory, and permitted them to come on board. They remained the whole day; but their conduct was altered very much for the better, particularly the old chief, who was now content to remain on the quarterdeck, to which they were restricted. I think the display of our cleaning arms before the chief on the day previous, led him to imagine we were intent on doing them mischief, and seemed to account for this sudden change in their manner.

The system of keeping them out of the ship except with permission, and then to a very limited number only, being once commenced, it was continued throughout our stay; and although it was very difficult to make them understand the necessity for it, which made it disliked, and was the cause of some ill-feeling towards us, it was impossible, through their numbers and want of honesty, to adopt any other course. At first we endeavoured to explain to them that we wished all to come on board in turn; but so far from entering into this view, those who were admitted and remained the whole day would invariably be the first alongside on the following morning, and be the most clamorous and least satisfied of those not admitted. These disappointments at not being allowed on board were retaliated in one or two instances by parties landing and carrying away our driftwood collected in a stack on the spit near the ship: this was found too laborious a revenge for them, and fire was tried; but a boat being sent, they pretended it was an accident, and did not repeat it.

On the occasion of our cutting into winter-quarters, our men being of necessity much spread about on the ice, and frequently surrounded by three times their number of natives, much caution was necessary to prevent the tools from being stolen, and many slight squabbles took place between our men and theirs in consequence of their playing them tricks and trying to trip them up. On these occasions they always selected those of our people who, from their appearance, were thought least likely to resent their jokes; but in some instances they found they had mistaken their men. Points of this sort were the most difficult for a commanding officer to deal with, as it was not possible for him to prevent the provocation, and, when not resented, the motives were misunderstood.

A more serious affair took place on board the ship. The officer in charge, Mr Hull, second master, in keeping back a large powerful man that attempted to force his way over the side, had a knife drawn on him by a friend of the other's on board the ship, who immediately called out for the women and children to retire. Mr Simpson, the surgeon, was standing near, and very soon produced before the man with the knife one of Colt's revolving pistols, and explained to him the use of its six charges, which had the effect of keeping them very quiet for the remainder of the day. I met the women and children retreating over the ice ahead of the ship, and thought something must have happened, although they told me they were going home to dance. A chief arriving at the same time reassured the retreating party, when we explained to them that if they used knives we must use guns, but otherwise we wished to be good friends. Similar squabbles took place frequently whilst our men were employed building the storehouse; knives were drawn as usual, and in two instances the women and children were sent away. This was a cause of a good deal of anxiety with me, as our men, being unarmed, were very much at their mercy under such circumstances; and in the event of arming them, more forbearance was necessary than some of them would have been found to possess, from the frequent provocations they had received in return for the usual kindness and good-nature that characterise seamen. On giving the subject every consideration, and seeing that it must excite an unpleasant feeling for our men to have knives pointed at them without a means of defence being at hand, the quartermaster of the watch, and two petty officers of the party working on shore, were armed with pistols, but properly cautioned not to produce them unless under circumstances of necessity, as I hoped the mere knowledge of their having them would be sufficient. Of this we soon had an instance. One of them played off one of their usual practical jokes on one of our men by kicking him in the back of the knees when carrying a spar, for which he was rewarded with a blow on the face; he then drew his knife, when the corporal of marines coming up, and being known to have a pistol, the offender ran away. These sort of annoyances continued as long as our men had work to do outside of the ship, and when the natives were collected in any numbers; the difference of character displayed by them when so, and the reverse, is worthy of remark. In the former case they are bolder and overbearing, and, when meeting with parties, gather round them, and, apparently in a half-playful way, commence shoving them about and feeling their clothes, when, if they fail in getting what they want given to them, they help themselves, and with their knives soon remove any buttons that happen to be bright. This was all done, and the offenders mixed up with the rest, enjoying the thing as a good joke, before our people could look round them. On the contrary, when they are in small numbers, they are not like the same people, but seem quiet, harmless, inoffensive, and obliging; but even while displaying these good qualities, should their numbers become increased, they lose no time in throwing off their assumed humility to join in any plunder going on.

In landing our provisions, I was particularly careful to point out to the chief and other leading men that nothing was going on shore the nature of which they could not see, except salt meat, which was really the case, and this I knew they would not eat if it was given them; and on the day that all was landed, and the house locked, I showed them the carronade pointed at it, and told them it was to keep thieves away, thinking that a show of preparation would have the effect of saving us from any attempt at robbery on their part; but I have every reason to believe that some of those on board at the time were leaders in breaking into it three nights afterwards, when fortunately three small sails (ship's) were the only things they succeeded in taking away. A case of flour contained in tins, belonging to the officers, had been opened; but not found to be tobacco, as anticipated, and not liking to go away empty-handed, they had taken the sails. I was quite unprepared for this theft, which was effected in the night, notwithstanding a strict watch had been kept from the ship, and the house was visited every hour,—as I had been told, by the officers of the ship acquainted with their character, they would never attempt anything greater than pilfering small things lying about. This there was no remedy for, except keeping a good look-out. Now they had commenced on a large scale, I had to consider the best mode of checking them before anything of a more serious nature should be attempted.

A slight show of firearms, in the way of intimidation, in all our former cases, had the effect of restoring the stolen articles; and with a view to the same effect, I had a small brass three-pounder mounted on a sledge, intending to threaten them with a visit if the sails were not returned. By the arrival of a native, who came every morning with dog's food, we were informed that during the time of sleep some people had committed the robbery, showing plainly that the affair was well known at the settlement. Our people had in the mean time tracked them on the snow to within a short distance of it, when the sails had been opened and most probably divided. About 9 A.M. the chief came down, assuming a very determined air, with his musket slung across his shoulders, to offer his assistance, and go with us for the recovery of our sails; but as he proposed leading us to Cape Smyth, where he said they had been taken, and stoutly denied their being at Point Barrow, his services were declined.

I must mention here that this was the common excuse with them; when anything was stolen, they invariably pointed to Cape Smyth, and said the things had been taken there. It became so well understood at last that no notice was taken of it, particularly in the present instance.

The chief, after some hesitation, came on board, when it was explained to him that we were quite aware where the sails were, and if they were not restored I should take the gun (which I showed him mounted) to their settlement to look for them; at the same time I thought the opportunity of having his musket in my power too good a one to be lost, and took possession of it, telling him that when he had brought back everything that had been stolen from us, it would be returned quite safe.

This appeared to place him in a serious difficulty, and after repeating the Cape Smyth story a good many times, he returned to the town, and we went on with our work as usual, intending to await the result of his interference. In about two hours he came again with some evasive story, that they were going to bring the sails down. He remained outside the ship evidently much disturbed, but not mistrustful; there were also a few others, women and children, and one sledge.

We now observed with our glasses an unusual stir at the settlement. In the first place, some women and children were seen moving across the bay to Cape Smyth; afterwards the men were seen advancing down towards the ship, in three single files, armed with their bows and arrows and quivers. I fancied at this time I saw spears also, but did not observe them afterwards. The leading men were discharging their arrows ahead of them as they advanced, picking them up again as they reached them, which satisfied me their visit was not friendly, and my mind was soon made up to keep them in check at the distance of musket-range by firing over their heads, wishing above all things to avoid taking a life, unless under some urgent necessity. Our small force, forty-one in all, was placed under command of the officers appointed to guard the gangways, poop, and forecastle; and previous to their getting within range, a blank charge was fired from our eighteen-pounder carronade and the three-pound brass gun, which had not the effect of dispersing them, as I expected; and when within musket-range we commenced firing over them from the forecastle. This had the effect of dispersing them under shelter of the spit about fifty yards from the ship's bows. At this time one of the chiefs, who had been on board frequently, and treated with every kindness, made a rush down ahead of the ship, followed at first by others; but when he found the balls whistling over his head he dropped on his face to avoid them, running a few paces closer to the ship, threw down his bow and quiver containing seventeen arrows, four of them with barbed iron heads. This man had become very unpopular with the crew from some uncivil acts of his; and I have been able to understand since, that although the order to fire over his head was carried out, this direction was very much infringed upon. A few now extended themselves under cover of the house, but as a constant fire was kept up in that direction, not many attempted to reach it; and a round-shot, being fired so as to graze it, had the effect of dislodging them. At this time a false alarm was given, that they were breaking down the house and carrying things away. I was on the forecastle, and on hearing the report ordered the man next me, a marine, to fire at a man then escaping from under cover of it, and from the sudden way he seemed to fall and kick out his legs, I thought he was killed. Immediately afterwards the report was found to be correct, and no more shots were fired at them: and I had the satisfaction to find out that the man fired at (the only instance) was not killed.

As the chief, who had been lying concealed under an ice hummock not far from the ship, and who, I suppose, now saw no chance of gaining an advantage over us with his numbers, showed himself, and beckoned them back in a most energetic manner, causing a general retreat, and as our masthead afforded a commanding view, I was glad to find that they were all able to use their legs quite as well going home as they did coming out.

Although this affair would give them a poor idea of us as marksmen, not appreciating our motives, I considered that some of them heard the ball sufficiently close to their ears not to wish for a repetition. Mr Simpson, the surgeon, counted seventy-one, and allows himself to have overlooked ten: he computes the number at eighty, besides several stragglers, a computation I consider as near as could be obtained.

The chief, with another man, stayed about the ship for some time; but as no compromise short of the immediate return of our stores was contemplated, he was not allowed to remain long. My having his gun was an advantage I could hardly have expected; and as its value to him was far greater than anything they had stolen from us, I was content to wait the result of his interference, in the mean time not allowing any of them within gunshot until everything was returned.

On the following day we had a pacific message, to the effect that they were all asleep that day, but on the next all our things would be brought down. They had not left the spot they had advanced to, when I took a party away from the ship to try the range of the gun in a sledge, and to find how it would answer. I was glad they saw us manœuvring it, and as they still remained after motioning them to go away, a musket was fired wide of them as a hint to be off and report what they had seen, which I hoped would have the effect of quickening their movements in returning our sails. At 7 A.M. on the following day, the chief, with seven natives and a sledge, brought down the sails,—a maintopmast stay-sail, and mizen trysail, and a boat's cover, all much worn, and of no importance to us; but the act was the same, and required checking, lest other things we might feel the loss of should be stolen also. I was told the party seemed in evident trepidation. The sails had been cut into several pieces, adapted in size for their u-mi-aks' sails, and had been served out amongst the party. This would account for the difficulty the chief had in getting them returned, without having a slight brush for them previously. All the pieces were most ingeniously drawn together by the women, who had been employed the previous day and night about them, which occasioned the day's delay in their being returned. As it was necessary, while we were settling matters, to have a full restitution of everything stolen from the ship and boats since our arrival, a careful inquiry was made to find out every missing article; and as these included almost all the iron-work of the launch, which had been cut almost to pieces, in the most vexatious way, while turned bottom up on the adjacent island, I was the more disposed to push this point. When the full extent was known, the chief was acquainted that everything must be returned previous to his getting his gun, or the natives allowed to come near the ship. He then left and returned next morning with every missing article, when his gun was restored to him, and the natives came about us as usual. I had the curiosity to examine the charge of his gun whilst it was in our hands, and found it as well loaded with ball as we could have done it ourselves, although he had previously told us he had no ammunition.

I made him a small present of tobacco for his trouble, as I believe he was not a partaker in the robbery, and I gave one of his wives a knife, as she had been very industrious in putting the sails together. He made us understand that he had been obliged to use his knife, as well as his authority, to compel some of the thieves to give up their share of the booty. I was glad to have got the upper hand of them without any further trouble on our part, as, independent of the more important motives before mentioned, our own travelling-parties might be seriously inconvenienced from being at variance with them.

Notwithstanding these considerations, it is most necessary, for our preservation with such a people, to establish respect from them by a moderate resistance upon any undue encroachment on their part.

Had we not been employed on a service essentially of peace, I should have taken a party up to their settlement in the way of retaliation, on finding they had broken into our store. I am not certain that it would not have been the better plan in the present case, as kindness and forbearance are not understood by them; particularly after being fired upon once or twice without receiving any injury, they are likely to form an erroneous opinion as to the power of fire-arms, many of the present party, including the chief, being the same who followed Commander Pullen so pertinaciously along the Return Reef of Sir John Franklin, when the system of avoiding firing at them was adopted until the last extremity, and with the same good fortune in not sacrificing any lives. As an instance of their ingratitude, I found many who were engaged in the robbery of the house were of those who had been allowed on board every day, and had received considerable presents with the view of making friends of them in the event of our requiring a kindness in return. Whilst our misunderstanding was unsettled, a further enclosure was marked out to include the ship, house, and observatory. Round this a stout hawser was supported on small triangles, and in no instance during the remainder of our stay was this boundary, which necessity had given us good excuse for establishing, allowed to be infringed upon. A few troublesome characters, such as will be found in all large communities, gave annoyance occasionally for a month or two, by shoving each other purposely inside the boundary, when, finding their tricks provoked hard blows, little more trouble was given in this way.

Our intercourse from this time went on more smoothly, and the Esquimaux, feeling themselves in error, seemed to make many friendly overtures to regain our esteem. Some of these consisted in bringing down their tambourines alongside, and dancing and singing in large numbers. In this they had in some degree anticipated my intentions, as the officers were at the same time engaged in printing a notice for the lower deck of a "Native Dance," intended to be given in three days' time, with the view of showing them we bore no ill-will, and wished for a friendly intercourse; and as it was to be the commencement of our winter festivities, and headed "Great Novelty," it had the desired effect of producing amusement amongst the crew.

At 4 P.M., October the 28th, our visitors were admitted to the number of seventy. After they were made to seat themselves round the deck, the entertainment commenced by serving each with a little tobacco; then our musical instruments (a violin, cornopean, drum, and triangle) played a lively air, which caused a general exclamation of wonder and pleasure, most of the party now hearing them for the first time. This was followed by a request for them to dance; and being supplied with a drum, they willingly complied. Our seamen danced in their turn; and in a little time the natives entered fully into the spirit of the amusement, stripping off their skin coats and dancing naked to the waist, with the temperature at 6°, showing the state of excitement they work themselves into: as the male performers shout in a wild triumphant manner, and all the lookers-on join in a chorus, and become as much excited as the performers, their appearance makes a scene as savage as can be well imagined. By 10 P.M. the party broke up, all appearing to have had dancing enough ; the whole company seemingly pleased with their evening's amusement. When we came to take down a few flags that were hung under the housing for ornament, it was vexing to find several large pieces cut out of them as if in handfuls. The chief and some others remaining appeared sorry, and promised the pieces should be returned, which was faithfully done the next morning.

On the following day I paid a visit to the village, accompanied by Mr Simpson, the surgeon. We were followed by several idlers from about the ship, who, as we neared the huts, spread the report of our arrival, which soon caused a great crowd to gather round us, following to the chief's hut, where we found him on his house-top ready to receive us. The winter huts were now covered with snow ; the chief's stood about five feet above the ground, with a square opening at one end, into which we followed through a low dark passage sloping downwards for five or six yards, when we stood beneath the opening in the floor of the inhabited part of the hut. It is circular in form, just large enough to admit one person at a time. Passing through it, we stood upon a smooth boarded floor, about 16 feet by 10 feet; the roof was seven feet high, and in the centre was a small square skylight, covered with transparent whale membrane.

The transition from the daylight and glare to the dark passage was sudden, and in some degree prepared our eyes for taking in at the first glance the appearance within. We were placed in the centre of the hut ; the chief, with a wife on each side of him, sat opposite to us. There were four or five young men, and two women with children, lying about the floor, all naked to the waist, the children perfectly so. The first breath of the interior was rather offensive, but we soon got accustomed to it ; and as the temperature was already high, being followed by a number of men whom it seemed impossible to accommodate in so small a space, it soon became insufferably hot, when the temperature was easily reduced by cutting a hole through the skylight, which made a very agreeable and necessary change in the air we were breathing.

Our visit seemed to give great satisfaction, and was commenced with a smoke, according to the method of doing everything. For the remainder of the time, about an hour, I endeavoured to find out from them how far the nearest winter settlement of natives was east of this place, with the view of communicating with them before the severity of the winter set in; but I could not make out anything very clear from them. We have a great difficulty in making them understand that our business is not bartering, as their ideas do not extend beyond that, although, from our constant repetition, they appear to be aware that we are waiting for two ships that have gone far away into the ice. The chief's was the only house we visited; and, returning across the bay to the ship, we were accompanied by a young man and a boy, who talked a great deal more than we could understand; but the former, in explaining to us the sort of tobacco that had been given him on board a ship, twisting his fingers together to describe American twist or negrohead, led us, in his description of the vessel, to believe it might have been the Investigator or Enterprise leaving the ice this last season, but we were afraid of giving way to our credulity in supposing it to be either of them. As they willingly accompanied us on board, I was glad to avail myself of Lieutenant Vernon's knowledge of the language to sift their story more thoroughly. He allowed them, most patiently, to describe all they had seen in their own way, and eventually ascertained that the ship they were on board of had diagonal decks, and an ice-chock larger than the Plover's. The illuminators in the deck, they remarked, were square. These are the points that seem to have caught their attention, and were sufficient to show that they had been on board one or other of the ships; but when the captain was described as wearing spectacles, Captain Collinson was identified. The remaining point of importance was, that she was gone to the eastward the summer before last (1851), agreeing with the time the Enterprise passed. In the spring of this year I stood on the point from which she had been boarded, with the native who gave this information. It is Cape Governor Simpson, and forms the western point of Smyth Bay, distant about forty-five miles to the eastward of Point Barrow. The Enterprise seems to have been delayed off it with light winds, but on a breeze springing up she was lost sight of to the eastward. Two u-mi-aks got alongside of her, and the people speak with pleasure of the presents they received: and it is worthy of notice that a particular kind of tobacco, with which we know the Enterprise to have been provided, led to a voluntary description of their having boarded a ship, affording more information in a few minutes than all our inquiries of the chiefs and others in several months had done.

After some experience with them, we found more information was obtained by casual observations of their own, brought about by something they might happen to see or have shown them than by asking direct questions, as it seems difficult to lead their thoughts from the passing events around them.

On the evening of the 5th November, the crew had a little recreation with the immolation of Guy Fawkes; and the natives, being told that he was a "big thief," were at great pains to get an opportunity of expressing individually that they were not thieves, which was very amusing, and seemed as if they expected to be treated similarly. The ceremony concluded with a rocket, on which they retired to a distance in dismay, and were evidently much impressed with the whole proceeding. They were afterwards gratified with a dance on board, which seemed to restore their confidence, and closed the amusements for the day.

As some of the crew had expressed a wish to see the Esquimaux village, and understanding there had been an invitation from one of the chiefs to witness a dance, I gave eight men leave on the afternoon of the 21st of November. All went well with them until one of our party, a quartermaster, who had been obliged to handle one of the natives rather roughly on his attempting to force his way into the observatory, was recognised by his old antagonist, who became furious with passion, and immediately attacked him, but was prevented by others from using his knife. He then endeavoured to get some of our people to his hut, who, seeing he had some weapon concealed in his sleeve, declined, when several of the Esquimaux hustled and tripped up our men, who still had protectors sufficient amongst the crowd to bring them away in safety. One man, the carpenter's mate, who by some means got left by his companions, was seized round the arms by two men, whilst the man who attacked the quartermaster picked his pockets of some tobacco and beads he had taken with him for the kind purpose of distributing amongst them. The chief was very much put out by the affair, and tried to get some of our people to his hut: his great aim being to get gunpowder, I suppose he fancied this display of feeling would not be favourable to his wishes.

After this attempt at a friendly intercourse, we all ceased going to the village for upwards of two months; when, seeing no inclination on the part of the men to renew their visit, I found if we pursued the system of holding back, a valuable opportunity would be lost of acquiring a knowledge of their habits which it would be interesting to become acquainted with, and for this purpose the officers again went occasionally to the settlement, taking care to go with some influential man, who kept away those inclined to be troublesome.

This was again very soon put a stop to by our not allowing the man to come to the ship who had behaved so badly when our men were at the huts; and in enforcing this measure we went near to having another affray with them. He had been to the ship several times since the occurrence took place, and on being told, always went away. On the occasion in question he evidently intended to remain and to get up a disturbance, if possible. The chief, who was on board, seeing this, went outside and tried to get him away, to no avail, and another chief tried also. The title of "chief" given here is merely nominal, as, in a community where every man has to provide for his own wants, the most industrious, bold, and successful hunter becomes, from the property he possesses, of more consideration than those not possessing those qualities; but this does not extend beyond his own boat's crew or hunting-party for the time being. Seeing this man was resolved to remain whether we pleased or not, I thought it necessary to have our own way as to who should visit the ship or who not; and as his conduct was well known to his countrymen, and condemned by many of them, his case was a good one to enforce. The great objection to any sort of disturbance was the difficulty of making the well-disposed understand our motives, or the separation of one from the rest as bad; and these cases rarely occurred, except when they were collected in large numbers, which made them imagine they could do as they pleased. While the altercation was going on outside, the natives on board, of whom there were several, began to leave the ship, although most of them had been visitors and declared friends, without, as it appeared, possessing any confidence in us. I then directed Lieutenant Vernon to go out and once more tell him to go away. The natives thinking he was the leader of an attacking party, two-thirds of the men, and all the women and children, walked straight away for the village, leaving the bad character in the minority, which he had sufficient tact to discover, and followed the others, when quietness was again restored; and those who had made such a hasty retreat out of the ship were now anxious to return again, but others were chosen in preference, from those who had remained as spectators, and were apparently indifferent. The crew were at dinner at the time, and they were not disturbed, as our state, with regard to arms, is one of constant readiness. The chief afterwards told me very impressively not to go to the village; a privation very little felt personally, but showing the absence of control on the part of the chiefs, when our being at variance with one individual of bad character put an end to friendly communication between us and a whole tribe.

He continued to make us visits at intervals of two or three days; but when, to put a stop to them, on one occasion Lieutenant Vernon had, with great good-nature and patience, walked with him a considerable distance from the ship, and left him to return, he followed him back leisurely, when I saw there was no means of dealing with him by kindness, and ordered two marines with muskets, outside on the ice, to prevent his further advance to the ship. On seeing them he seemed to fly into a great state of excitement, and presented his breast to be fired at, when one or two well-disposed individuals took him away, and he did not repeat his visits again for some time; when, being told by the chiefs that he was now good, I was glad of an opportunity to come to terms with him, seeing that disputes would do us no good, however much we had right on our side; and as I thought we had established our point of keeping him away, to as great an extent as could be expected with a savage, it was explained to him by one of the chiefs that he was to go away, and when we intended that he should remain, a message would be sent to him. Understanding this, he went away much pleased, and on being sent to in a few days, he came down in his best clothes and behaviour, and was allowed to remain, which put an end to all disputes between us.

During this day, which was fine and clear, with a temperature at 23°, we had eighteen natives on board, and seventy-two were counted outside, or in sight of the ship, making ninety in all, men, women, and children, a number not unfrequent in the depth of winter when they were not engaged hunting. No temperature was too severe to keep them away; with the thermometer at 30° and 40° below zero, they commenced arriving alongside as early as 6 A.M., three or four hours before daylight, and those not admitted on board sat on the snow, laughing and playing about as cheerfully as we should have done in sunshine. A party of six at this time had a narrow escape from being starved on the ice, on which they were adrift for six days. They went out to hunt the white bear, when the floe became detached, and drifted into the open water. They were saved from this perilous situation by the chance of its being again brought in contact with the land floe. Although the weather was very severe, so perfect is their clothing that the only injury they received was some frost-bites about the face: as they were without food, a description of their sufferings would be one of thrilling interest, if we could understand their language sufficiently to appreciate it.

A short time afterwards, on a general break-up of the ice, one man was carried away and lost. Finding he had left a wife and two children, I sent a message to her to come to the ship, and on making her some useful presents I endeavoured to have explained to a party assembled, that our business here was looking after people who were also in the ice, hoping it would have a tendency to gain their assistance and friendship in favour of any of our parties travelling along the coasts they frequent.

Several stratagems were tried to find out if we were on the look-out at night. Small parties, generally two or three women, came alongside, and on being sent away would make some excuse of going out sealing, or some such pretence, to keep in our neighbourhood; but our watch was always too vigilant for them. One man was caught, in the middle watch, coming out of a tent erected over a theodolite on the ice close to the ship, and was brought on board a prisoner. As he had not been able to steal anything, the case was not strong enough to inflict any punishment on him, and he was permitted to go away the following day at noon, with a promise of a good thrashing if he was caught again. While he was on board, only four or five women and the chief came near us. I was glad to see the latter, as he witnessed our leniency on the occasion; but he seemed indifferent as to whether we had punished him or not. This detection put a stop to their nightly excursions, and left us unmolested.

A few of them from the first showed disinterested partiality to us, which was appreciated, and they were invited to sleep on board on a few rare occasions. For some time they were nervous about it, but mustered courage enough after a little time. My object in allowing them to do so was to give them confidence, and let them see some more of our habits, which I thought would raise us in their estimation.

In the early part of February we were startled by a native report that a large ship with a great many men was wintering to the southward at Point Hope. A party belonging to Cape Smyth had lately returned from Point Hope (a journey frequently made by them in the winter), and had, it was said, been on board the ship; but knowing it to be an impossibility for a ship to winter there, not much attention was paid to it, beyond thinking there generally exists some grounds for such reports, and when well sifted they can sometimes be traced out. The present report seemed to be a forerunner of a party of natives belonging to Point Hope, who arrived at Point Barrow about this time, and soon afterwards visited the ship. The information received from them was of there being a large ship somewhere to the south, with very little men on board; whether we mistook their "little" for few—or most likely it was one of many such stories these people have amongst them, from the habit of repeating them one to the other without reference to the date, as in this case, if the story mentioned was traced to its source, it would most likely prove to be some whaler visited by them in the summer.

This circumstance is mentioned to account for the way reports are carried along a great extent of coast by natives meeting for a short time for the purpose of barter, and at considerable distances from their respective settlements, when, as may naturally be supposed, an interchange of news takes place, and thus becomes extended onwards, subject to such change of version as the repeating tribes are likely to give it, making it necessary to consider their reports well before taking much notice of them unless accompanied by some token of their authority.

The chief of this party was a pleasing, spirited man, about thirty-five years of age, and was accompanied by his two wives, who were good-looking young women. He had come on a bartering expedition to dispose of some copper kettles, receiving glutton skins in exchange. He recognised Mr Simpson, the surgeon, as having seen him at Hotham's Inlet, and he was also familiar with the name of Captain Moore and some of the former officers of the ship. He described his journey to us, of which I was anxious to get the particulars. He had slept fifteen times, eight of them on the snow, but he showed no signs of having suffered from the cold. He described the sea as being open all along the coast to the southward, but he was not familiar with ships, which I accounted for by the lowness of the land about Point Hope deterring the whalers from closing it, otherwise there are many in that latitude in the summer. He seemed to be a poet, and favoured us with an extempore song of some length, which included the name of the ship and some of our own. He afterwards passed his hand down his stomach several times, expressive of great friendship, and then fixed his forehead against mine, and used it as a fulcrum to rub noses several times, a ceremony not very agreeable in his heated state from singing. A chief of this place was of the party, which made him feel quite at his ease, otherwise they are mistrustful on the first visit. He described his people's barter with the Asiatics of the opposite coast (the Tchuk-chi, or, as they call it, Tsau-chu) to consist of the martin (sable), fox, wolverine, wolf, and bear skins, and sometimes whale-oil and fish; for which they get in exchange kettles, tobacco, beads, knives (Russian), and walrus teeth. While he remained he was a welcome and constant visitor. The southern natives are far more agreeable than those to the northward, who have been rarely visited; the officers speak of the former as more grateful and modest, particularly the females, while the latter are thought impudent and ungrateful. On leaving, he sung of the good name he should give the ship on his way down the coast, which will be of benefit to our boats on their expedition to Cape Lisburne.

About the middle of February a great many of the natives removed into the interior to hunt the reindeer, found very numerous on these vast plains, which form the north-western termination of America. As I was curious to see something of the country, and their mode of living at this inclement season, I thought my making an excursion to their hunting-grounds would show we placed every confidence in them. I accordingly left the ship on the 1st March, accompanied by Mr Gordon, mate, two seamen, with a native guide, and sledge drawn by six dogs, carrying tents, guns, and provisions. Our route into the interior was S.S.W. (true), across snowy plains, differing little in appearance from the bay ice, except on the approach to a lake, when the rise became more sudden, and being exposed to the sweeping winds, grass was occasionally seen in spots through the snow, offering, with deer, of which we saw a great many, the only thing worthy of notice. Three days and a half heavy travelling brought us to an encampment on the bank of a river. The people had a hole through the ice, which we found to be seven feet thick, and reaching within one of the bottom. Their houses differed from those described by Captain Parry, in being excavated in the snow lodged on the river, the ice of which formed a perfectly even floor. Their position from outside was only observable by seeing the implements of chase belonging to the owners in a group over the top. We found them in no instance wanting in kindness, but their character for begging had not fallen off since leaving Point Barrow. Their mode of killing the reindeer is novel, and such as nature has pointed out. The country is so open that they have no means of approaching the animals under any cover; they therefore dig deep pits in the snowy ravines, selecting places where the surface is even, to cover them lightly over with slabs of snow. The moment the animal puts his feet upon them he is precipitated into the pit, the depth of which is too great for him to leap from.

Having effected my purpose of visiting and going amongst them, I returned to the ship after an absence of seven days. The position of the place by astronomical observations was S. 40, W. 38 miles from the ship.

Nothing further worthy of remark occurred in our intercourse, until I was setting out on a journey along the coast to the eastward, when some of the worst-disposed tried to deter a young man from accompanying me as guide by threatening to follow and murder us when we slept, the guide amongst the number, upon whom it appeared to have no effect, beyond his repeating the story and advising that the men should not be allowed into the ship during my absence. This report made it evident that the fact of our force being divided had been talked of, and I felt sorry our residence amongst them for so long had produced no better feeling; but I considered it necessary to show we were capable of defending ourselves, and travelling when we found it necessary. If this was not established, our position, confined to the ship by a tribe of unarmed savages, would not be very flattering. With these feelings I set out on my journey, well satisfied with the resources of my party, and with the judgment and discretion of the officer (Lieutenant Vernon) in whose charge the Plover was left. On my return after an absence of twenty-five days, I was glad to find everything going on as favourably as possible. The natives were returning from the interior in large numbers, and the day previous, 27th of April, as many as 40 sledges with 93 people crossed the bay to the village from the hunting-grounds.

From this time we procured an abundant supply of venison, the only instance during our stay of our receiving any benefit from our proximity to them. It lasted nearly two months, and had the effect of restoring for a time the health of the crew, previously much debilitated and exposed to scurvy.

The appearance of this disease was attributable to the great inequality in the character of some of the provisions in the older supplies of preserved meats and flour; and it became necessary to discontinue the use of beef altogether, and to increase the issue of the last supply of preserved meats and preserved potatoes.

The season for their whale-fishing was now approaching, and commenced on the 7th of May; the distance to the open water was about four miles W. (true) of Point Barrow. On the 11th, hearing they had caught a whale, I made an excursion to the scene, with the hope of being in time to see its distribution; but on arriving, the only remains we could discover was about half a pound of blubber, to so good an account do they seem to turn every part of the animal. We found the open water extending E.NE. and W.S.W. (true), and no bottom with ten fathoms. The ice to the southward seemed open, and I supposed the water was free in that direction to Behring Strait. It would be interesting to be able to conjecture its north-eastward extent, the wind blowing, it may be said, constantly from that quarter; if the ice should happen to be broken, it must clear a large space of water.

This pursuit occupied them until the 21st of June, when most of their u-mi-aks were brought to the land to be prepared for their summer journey to the eastward, previous to which they seem to allow themselves ten days' enjoyment, which is passed in eating, smoking, and dancing; they then commence the work of preparation.

Whilst treating on this subject, it is worthy of being mentioned that the natives state that the whales make their appearance off Point Hope in April and May, when the ice there breaks up into fields, and that most of them have disappeared by the time the ships arrive. About the same time these animals also appear at this place, and are pursued by the natives in their u-mi-aks, as mentioned before, until June, when few are seen, and in July none in this neighbourhood; and the people believe they retire northward, to return this way in August and September. The masters of whaling vessels have informed me that whales are less abundant in the open water in July and August than in September.

A month previous to this time we received, quite by accident, some very useful information. Two of the officers, Lieut. Vernon and Mr Simpson (surgeon), conversing with one of the chiefs more intelligent and communicative than the rest, asked him if he had ever seen any boats like ours along the coast, when he said he had, near the Colville river. Finding that Mr Simpson, by turning over the leaves of Commander Pullen's journal, was able to describe the occurrence that took place, the man unwittingly gave a detail of the whole affair, which corresponded with the written accounts, even to the wind; by which it appears the chief of this place, with his Hudson Bay gun—Barnett, 1843 (*vide* evidence taken before the Arctic Committee, questions 1384-5-6), —is the individual described by Commander Pullen, and he appears to have taken a large number of men with him to follow the boats. The man afterwards seemed very uneasy, and was perhaps thrown off his guard when he told so much, but the officers supposed that he thought if the book told so much, they must also be acquainted with every circumstance by the same means, and that he was doing no harm in acknowledging the facts. He was very anxious from the first to learn the names of Commander Pullen and his companion Lieutenant Hooper (a prevalent habit with them), but he was not told until he had identified them in the most minute way. In further conversations connected with the foregoing circumstance, we were enabled to identify this people as the Western Esquimaux, and the same man traced out for us the different stages of the journey. This point had been of doubt with us; supposing the distance from Point Barrow to Barter Island 240 miles of an exposed coast, it could not be accomplished by them in the short season of open water, as their skin boats, when deeply laden, are not adapted for a seaway. These difficulties were explained by saying they take the boats away on sledges a month before the ice breaks up; the exposed parts of the coast and large bays are avoided by an inland navigation through rivers and lakes of which we were ignorant.

The first stage of the journey is the Colville, and occupies ten days, where they meet a friendly tribe of Esquimaux, called Nuna-tag-miutes (supposed by Messrs Dease and Simpson to be a name for the Russians), but there is no doubt of their being Esquimaux, and wearing the lip ornament peculiar to them. They confine themselves to the rivers and land from which their name is derived, "Nuna" (land), and have a communication through the rivers of the interior to the coast about Hotham Inlet, as our informant told us he had met a woman two successive seasons at the Colville, who had been frequently aboard this ship in Kotzebue Sound in the winter of 1849, and had travelled from there through the interior without having touched upon the intermediate coast.

The journey to the Colville is anticipated by them with

much pleasure, particularly this year, as they frequently spoke of telling of the wonders seen on board this ship, in addition to the pleasing thoughts of feasting and dancing they were to enjoy with their friends. From thence a select party extend themselves to Barter Island, the women accompanying them to within a day's journey of meeting the eastern people, when the men advance and conclude their business as expeditiously as possible. They give an amusing account of the mutual distrust that exists between the two parties. The western people never sleep while they are in sight of the eastern, and all bargains seem to be made knife in hand. The articles and system of bartering appear to be as described by Sir John Franklin, but I think later writers on the subject have supposed that Russian goods find their way to the north coast from posts on the Colville, which we have not been able to verify. This tribe receives Russian (Siberian) articles from the Point Hope people, previously alluded to, which they convey to the eastward and exchange for English knives procured from the Hudson Bay posts; but there is no intermediate supply from the Russians that we could hear of. The journey east from the Colville takes them ten days, which they describe as being made always against the wind; and the return to Point Barrow from Point Berens occupies little more than two, while they sleep in their boats, and allow them to drift before the wind; this would make it appear that easterly winds must be very prevalent in the early part of August. Mr Simpson, who has taken much pains to inform himself on the subject, considers the 25th of July the time of their departure from the Colville: this has been further established by the fact of the party who attacked Commander Pullen on the 9th of August at Return Reef being then on their return from Barter Island. The time of their annual visit never differs probably three days, as we have found them generally as accurate in that way as if they were acquainted with dates. From this information it may be concluded that the Esquimaux make one-third of their summer journeys by carrying their boats over the ice—before a ship or her boats become available, which gives them a season so much longer, at a time when the weather is very favourable; this will have been pointed out to them by necessity, from the shortness of the season of open water, which, if they confined themselves to it, would limit their excursions to a very short range. Their return for the winter takes place, as we have seen, about the 10th of September, and seems to end their labours for the year.

Two days after receiving the above information, May 20, a native man was observed outside the ship wearing a small canvass bag with an address written on it—"To the Chief Trader of the Russian settlement, N. America,"—which caused considerable curiosity on our parts; and on making further inquiries he told us he had the paper in his hut that was originally in it; on which he was despatched with the promise of a large reward of tobacco if he brought it to the ship. Some hours afterwards he returned with two torn pieces of paper, complaining that his little girl had destroyed the rest; fortunately, the most important part was preserved which has been enclosed. This information is so far satisfactory, as it shows Commander M'Clure to have been making his passage along shore, which has been further confirmed by natives here, who were on board the Investigator at Point Berens or Return Reef, where they describe her to have had an easterly wind with no ice in sight. The Enterprise, it has been mentioned, was visited the year following not so far to the eastward by eighty miles. From these positions it is reasonable to infer they had not again communicated with the shore to the westward of the Mackenzie, as the natives of this place, who are in yearly communication with the Esquimaux who frequent the mouth of that river, have, to our repeated inquiries as to whether they had heard if the ships had been seen by the other people, answered they have not, or they would have heard of it. It will give some idea of the difficulty of making the Esquimaux comprehend the meaning of our notions, when it is considered the Investigator had an interpreter on board, who will have explained the object of the letter given by Commander M'Clure, and we had been eight months constantly making inquiries for such things, and endeavouring to explain our business here, without the individual charged with the paper knowing what it was for; and if the canvass bag had not been found useful, its contents would never have been known.

In order to impress upon them the value of such things as papers and messages, I gave the man a considerable present of tobacco, very much to his surprise and that of his companions, which had the effect of producing an old American song-book, the only article of paper remaining in their possession.

I have now to mention the occurrence of a native man being shot by accident, which at the time caused us considerable trouble and uneasiness, but eventually I have had reason to believe it was viewed by most of the Esquimaux in its true light.

On the morning in question, June 8, the quartermaster of the watch, David Dunstall, came into my cabin, and informed me he had had a dreadful misfortune, and, to my horror, that he had shot a native alongside the ship; and on hurrying outside I found the man was shot through the head, and must have died instantaneously. The man who had been the cause of the unlooked-for event showed by his manner that it had been an accident; and, upon making some further inquiries, I found that several natives had arrived alongside the ship previous to the time they were allowed, and although desired on that account to go away several times, they could not be induced to do so, and the quartermaster of the watch took out a fowling-piece in his hand, in order to frighten them, and when motioning with it for them to go away, it went off and lodged the contents of the barrel in the back part of the poor man's head. The remainder of the party, five or six, ran away so speedily that there was no means of overtaking them, and the body being left, it became necessary to consider the best means of disposing of it. We soon afterwards removed it to such a distance from the ship that the natives could advance to it without fear of us, and at the same time it gave no pretext for their coming any nearer to the ship. When this had been done, and a large quantity of tobacco left with the body as an intimation of our friendship, all that we could do was to hope that some of our friends amongst them would still have sufficient confidence to come down and give us an opportunity of explaining this affair. In this expectation we were not disappointed, as two of the chief men came to the ship at once, having, before leaving the settlement, exerted their influence to quiet the people in their first outbreak. One of these men, who was remarkable for his intelligence, was made to comprehend the possibility of such an accident; and great pains were taken to show him that the charge was shot intended for birds, not men. When this impression was established, we requested them to go back and explain it to the people. By the time they had arrived at the place where the body had been left, a great many had collected, amongst them the wife and friends of the unfortunate deceased, but who, I was glad to find, had left no children. They sat round, and appeared deeply engaged in conversation for about two hours, listening, as we supposed, to the explanation; then they seemed to examine the body, and his own deer-skins having been brought down, he was wrapped in them and placed on a sledge, which was drawn by his wife leading and four men, one following, across the bay to the cemetery, near Point Barrow. None of the others accompanied the procession. A few of them came as usual alongside the ship, but as they were for the most part of those known not to be friendly to us, our people were kept on board to avoid any treacherous retaliation. During the day I was gratified to find the wives of the principal chiefs came on board, and expressed their sorrow at the absence of their husbands, who were at the open water looking after whale; but they had been sent for, and were coming on board as soon as they returned. They told us also that all work was stopped for five days, the women not being allowed to sew for that time, which seems to be a general custom on the occasion of any deaths, and remarked that we ought not to have any hammering on board for the same time; and as I was anxious to show every sympathy in our power, the caulkers at work outside were ordered to cease work, and the ensign hoisted half-mast, the meaning of which was explained and understood by them. In the evening one of the chiefs who had been first down after the accident visited us with his wife, and brought the intelligence that a division existed in camp as to revenging themselves on us, but, as the chiefs were unwilling to favour it, it seemed probable nothing of the sort would be attempted.

However, proper precautions were taken to avoid any surprise, particularly as a thick fog at the time of sleep favoured such a design.

On the following day the four chiefs, with their wives, came on board, when, with the assistance of the officers, the whole affair was again explained, which they seemed to comprehend, and appeared to entertain no fear of going below as usual when asked. We were quite aware of their inability to control individuals, but they have to a certain degree a good deal of sway, and recommended us to make presents to all the people who lived in the same house as the deceased, and cautioned us not to stray away from the neighbourhood of the ship, as the feeling of a great many of them was uncertain. On their leaving they were made presents, and desired to bring the house-party to the ship at the expiration of five days, the time they strictly confine themselves to the house, and, I believe, are seldom intruded upon. They arrived at the ship in due time, numbering as many as ten, accompanied by the chiefs. The widow was young, and seemed in unaffected grief, which the large presents she received did not dissipate; but time and a continual course of kindness on our parts brought her to forget her loss so far that she regretted to me, and I believe in sincerity, the thoughts of the ship going away. Our intercourse from this time, as might be expected, could not be conducted with the same satisfaction it had previously been, although no trouble or inconvenience was spared to do away with their unfavourable impression. The views they entertained of the case very much assisted a reconciliation, as they had no idea of any such thing as a general control existing amongst us; the act was set down as purely an individual one, and the whole fault of it attributed to the man as an act of his own, while they seemed to consider the remainder not in any way concerned in it.

About fourteen days after the occurrence a large collection of people had assembled at Point Barrow to witness some dancing festival previous to setting out to the eastward.

A party from Cape Smyth afterwards endeavoured to get up a united force to attack the ship, without being able to succeed. However, the chief and a good many of the Point Barrow people wishing to make a merit, for which to be rewarded, of their not joining our enemies, I was obliged to tell them that I should be very sorry if the people came down with bows to the ship again, but if they did, a good many would be killed. This appears to have put an end to it, as we were next told of the departure of the hostile party, and were not troubled with any more reports of that description.

We now watched with some anxiety the preparations for the departure of the natives, as the time was approaching for the departure of our boats for Cape Lisburne, and their transit over the ice could not be effected without some risk of interruption until the numbers of the natives became decreased, so that we were much pleased to see them in a state of forwardness. On July 4 they commenced leaving, and continued to follow for three successive days.

Each party halted to sleep near the ship to have an opportunity of begging from us till the last moment, and as I had prepared some printed papers for them to distribute among the eastern Esquimaux at Barter Island, they were treated with every indulgence; and those intrusted with the printed papers received a present of tobacco, some buttons, prepared in England, bearing information concerning the arctic search, and some other small trifles to keep them reminded of their charge, which they faithfully promised to pass on. I redeemed a long-made promise to the chief, giving him some gunpowder. I think it had a good effect, by showing we had no fear of them; and I was fully aware he had a sufficient quantity for mischievous purposes, as his gun, when in our possession, we found well loaded with ball-cartridge.

I accompanied one party a short distance on their journey to see if I could gain any useful hints for our own boats. The u-mi-aks were secured on small sledges, and seemed to be easily drawn by three people. The principal part of their barter (whale-blubber and seal-oil) was carried on small truck-sledges, drawn by the women and dogs, the men seeming to confine themselves to the charge of the boats; but at lanes of water and different passes they mutually assist each other. The rate of travelling whilst moving was expeditious, but they made frequent halts to smoke; and before I left them, although near their place of encampment, they ate a hearty meal.

Between the 4th and 7th of July, as many as twenty-seven to thirty u-mi-aks, accompanied by 150 people, passed to the eastward, much to our satisfaction, as it left us with comparatively few to trouble us in our preparations for sending away the boats.

On the 7th I made an excursion across the bay to observe the state of the sea-ice, and walked two miles to the westward in the offing without being able to get a glimpse of open water in any direction from the highest hummock. This made me form the idea of taking the boats over the ice to the southward until we fell in with it.

On the morning of the 9th I left the ship with the gig and whale-boat intended for Cape Lisburne. They were placed on two strong sledges drawn by their own crews and officers, Lieutenant Vernon and Mr Gordon (mate), assisted by an auxiliary party of ten, including myself and the carpenter, making up twenty in all. Thirty-four days' provisions for the boats' crews, clothing, ammunition, &c., were carried on two native sledges drawn by dogs, a third conveying provisions for the auxiliary party. Parts of the transit over the ice were attended with difficulty, and the whole was very laborious, requiring the utmost strength of all the parties, assisted at times by natives, and for two days by a strong and favourable breeze, enabling us to make sail on the boats, which lightened the labour very much. Travelling to the southward for three days, the open water appeared within two miles of the beach at what seemed a most eligible place for launching the boats. I walked, accompanied by Lieutenant Vernon, to the open water, but the hummocks intervening seemed to offer too great an obstacle to the transit; however, on the following morning, July 12, as the weather was favourable, I decided to attempt it, and succeeded in launching the boats safely at noon. They soon after shoved off with a favourable breeze, which they held for eight hours only.

On the 15th, whilst calculating on the progress they might have made, a party of people were observed coming down at 8 o'clock in the evening, tracking an u-mi-ak, and on making out some of our people amongst them, our astonishment was so great we could hardly believe our eyes. I went without loss of time to meet them, not in a very agreeable state of mind; but when I counted the full number of the party I was comparatively easy. It appears that on the night of the 13th, finding the ice closing on them, they managed to get the boats on to the floe, and fancied themselves safe, although drifting to the northward. The ice continued to press in towards the land, crushing the floe, and piling it up to nearly twenty feet in height; and at length the floe gave way beneath them, and the gig was partly filled with pieces, and could not be withdrawn further towards the land. The whaler, a lighter boat, was also stove; and as she could not be brought further, the whole party were obliged to make the best of their way to the shore before the ice should ease off, as it seemed inclined to do, and leave them on detached pieces. Fortunately the danger, so imminent, was of that slow, though overwhelming nature, that they had time to provide themselves with three days' preserved meats, and their arms and ammunition, with which they were enabled to reach the ship. I have forwarded Lieutenant Vernon's journal, which will show their Lordships how that officer and his companion, Mr Gordon, acquitted themselves on the occasion; and I have much satisfaction in mentioning the high terms in which he spoke of the conduct of the crews, not a man of whom expressed his opinion until they had decided on deserting the boats, when, from their manner, he concludes they thought it high time; and he further describes them as obedient, cool, and intrepid from the commencement of the danger until their return on board. The loss of the ship's boats was a heavy misfortune; but when it was considered the whole party had found their way safely to the ship, we were too much rejoiced to give a look of regret after the boats, which could not have been saved.

Two days after the return of the boat-party I had serious thoughts of despatching another in a native u-mi-ak, to keep our appointment at Cape Lisburne; but on weighing the matter maturely, I gave up the idea, as by no means in our power could it reach the rendezvous by the appointed time; and as the ship would be obliged to wait for the return of the party, the delay would render it impossible to regain this place as a winter station, in the event of its being so ordered. With these considerations, and a present prospect of an early release, I thought it better not to divide our small force at a time when the services of every one would be required to extricate the ship, and to take the earliest opportunity of going to the south in order to recruit the health of the ship's company.

As early as July 25 we were enabled to swing the ship for local attraction; and the ice having cleared away for a considerable space in our neighbourhood, we moved into the fairway to be ready for the first opening in the offing.

The natives brought us a report this morning that our deserted boats had been seen drifted near the shore at the Sea Horse Islands, and their contents made a prize of by a small party, who succeeded in bringing the gig to the land. On our passage down, beating close along shore, some natives brought the shell of the boat off, exchanging her readily for a native u-mi-ak we had provided ourselves with in the mean time.

In watching the dissolution of the bay ice, we had flattered ourselves this season was fourteen days before the last, which it really was in the thinness of the ice; but from the absence of strong winds it remained stationary, at a time when the sea was open the preceding year. Every day at this period (the end of July) seemed to effect a marked change in the ice, and on the 30th a perceptible motion to the northward was observed in some of the heavy hummocks in the offing, and a lane of water sufficient for the ship was found by Mr Hull (second master), whom I sent to observe the state of the ice to seaward.

In consequence of this report I set out in a boat on the following day to trace its extent, preparatory to moving the ship, and to sound along shore to find how close we could approach in case of necessity. Everything seemed favourable to our purpose, and I fully expected to return in the evening and get under way; but on approaching Cape Smyth, I had the mortification to find the lane of water terminate so close to the shore that a native u-mi-ak could not have passed. This was one of the disappointments of ice navigation, which teaches one patience, and requires to be borne with, in addition to others we are liable to, even under favourable circumstances.

From the 1st to the 6th of August an officer was sent each day to Cape Smyth, to bring the earliest intelligence of any change in our favour; but each day brought the same unwelcome report, that the ice was still close with the shore, and no open water to be seen. It seems a strange complaint in these latitudes, but the fineness of the present season was the cause of our detention. During the month of July the wind was registered one day from five to six in force; for four days four; and the remaining twenty-six barely averaged two, proving the extreme of fine weather to be a greater obstacle in moving masses of ice than the reverse state of it.

On the 7th a fresh breeze from the eastward gave me some hopes of a change in our favour; and on visiting Cape Smyth I observed a narrow lane of water extending as far as the eye could reach to the southward. On getting on board at 8 P.M., we left our anchorage, where we had found shelter for eleven months and four days, seven days later in the season than Captain Parry had been enabled to leave Melville Island. We carried a favourable breeze for eight hours, and had made considerable progress, when the wind shifted to S.W., with thick weather and heavy rain, which made it difficult to avoid coming in contact with the ice, from the narrowness of the channel we had to beat in. A continuation of thick weather and light contrary winds on the following day prevented our making any progress; and in the afternoon, finding we were losing ground, I made fast to a large floe-piece, when we found the drift to be one mile an hour, east, being very nearly in the opposite direction to our course. After a few hours we were enabled to cast off, on a light wind springing up from the northward, and run along shore to the S.W.; but the ice had become gradually closer during the few hours of westerly winds, and we had much difficulty in picking our way during the night, as a thick fog prevailed. On the following morning, 9th August, the outer edge of the land-floe in the depth of Peard Bay was found to extend unbroken from the land to a distance of nine miles, and the ice outside seemed too close to offer us a passage in the desired direction; but as the wind was fresh and favourable, the ship was forced into open water by 10 A.M., having sustained some severe but unavoidable shocks in doing so. I was now anxious to keep close in with the land, in order to stop any boats that might be making their way to Point Barrow, in consequence of our failing to reach Cape Lisburne; and at 2 P.M., in passing the Sea Horse Islands, we got into three fathoms in endeavouring to close the shore for the purpose of sending a boat to put up a mark, which we were not able to accomplish, as the wind was found to blow rather on shore, and was too strong to admit of the ship lying off and on with safety. I accordingly bore up to run for Cape Lisburne, and on the following day, 10th August, at 11 A.M., communicated with her Majesty's ship Amphitrite, Captain Frederick, from whom I received their Lordships' orders, directing me to remain at Point Barrow; but the health of the crew would have prevented my doing so had I received the order previous to leaving; and as there seemed no difficulty in returning during the present season, I repaired, in company with the Amphitrite, to Port Clarence, to await the arrival of the Rattlesnake, and to obtain the changes found necessary in the crew, with an increased supply of provisions and fuel for another year.

Our return seemed the more necessary, as their Lordships' instructions direct provisions to be left in the neighbourhood of Point Barrow, which I had not done, considering at the time I left it was probable I should return. A reference to Sir E. Belcher's instructions relative to the north coast, seemed also to make it desirable that the Plover should return to the station she was known to occupy by that officer, in the event of any of his parties advancing with the certain hope of succour at Point Barrow.

ROCHFORT MAGUIRE, *Commander*,
Her Majesty's Ship Plover.

PORT CLARENCE, 21*st August* 1853.

THE END.

PRINTED BY WILLIAM BLACKWOOD AND SONS, EDINBURGH.

Printed in Dunstable, United Kingdom

91226148R00219